Springer Tracts in Mechanical Engineering

Springer Tracts in Mechanical Engineering (STME) publishes the latest developments in Mechanical Engineering - quickly, informally and with high quality. The intent is to cover all the main branches of mechanical engineering, both theoretical and applied, including:

- Engineering Design
- Machinery and Machine Elements
- Mechanical Structures and Stress Analysis
- Automotive Engineering
- Engine Technology
- Aerospace Technology and Astronautics
- Nanotechnology and Microengineering
- Control, Robotics, Mechatronics
- MEMS
- Theoretical and Applied Mechanics
- Dynamical Systems, Control
- Fluids Mechanics
- Engineering Thermodynamics, Heat and Mass Transfer
- Manufacturing
- Precision Engineering, Instrumentation, Measurement
- Materials Engineering
- Tribology and Surface Technology

Within the scope of the series are monographs, professional books or graduate textbooks, edited volumes as well as outstanding PhD theses and books purposely devoted to support education in mechanical engineering at graduate and post-graduate levels.

Indexed by SCOPUS, zbMATH, SCImago.

Please check our Lecture Notes in Mechanical Engineering at https://link.springer.com/bookseries/11236 if you are interested in conference proceedings.

To submit a proposal or for further inquiries, please contact the Springer Editor **in your region**:

Ms. Ella Zhang (China)
Email: ella.zhang@springernature.com
Priya Vyas (India)
Email: priya.vyas@springer.com
Dr. Leontina Di Cecco (All other countries)
Email: leontina.dicecco@springer.com

All books published in the series are submitted for consideration in Web of Science.

D. R. Raghavendra

Electrohydraulic Servo Systems

Applications, Design and Control

Springer

D. R. Raghavendra
Research and Testing
Department
Central Manufacturing
Technology Institute
Bengaluru, Karnataka, India

Founder Member of Fluid Power
Society of India
Bengaluru, Karnataka, India

ISSN 2195-9862 ISSN 2195-9870 (electronic)
Springer Tracts in Mechanical Engineering
ISBN 978-981-19-8064-0 ISBN 978-981-19-8065-7 (eBook)
https://doi.org/10.1007/978-981-19-8065-7

This Springer imprint is published by the registered company Springer Nature Singapore Pte Ltd.
The registered company address is: 152 Beach Road, #21-01/04 Gateway East, Singapore 189721, Singapore

TO
MY WIFE, KASTURI
and our Family

Foreword I

Ever since Joseph Bramah invented the hydraulic press in 1795, hydraulic power transmission has proved an extraordinarily successful means of controlling motion and force in applications that benefit human kind. To quote Bramah's patent, hydraulics is about using "dense fluids to operate mechanical apparatus's so as to act with immense accumulated force, [and] to communicate motion and power from one part of a machine to some other part of the same machine". As true today as in Bramah's time, hydraulic power transmission provides a unique ability to exert high forces at high speed, while maintaining an extraordinary degree of precision and control. What is more, at the point of power application hydraulic actuators (typically cylinders) are exceptionally light and compact, can be mounted remotely and moveably relative to the hydraulic power generator (power pack) and have proved to be remarkably robust and reliable.

A system which exhibits accurate control is one in which the physical variables of interest—which might be one or more position, velocity, acceleration, or force—can be precisely held at a desired constant value, or can be changed very rapidly to track a varying desired value, and are also barely affected by unwanted external disturbances. To achieve the most accurate hydraulic control systems, an in-depth understanding of the system's characteristics is required. Even when open loop control is possible, closed loop control will almost always provide far superior performance, but understanding the subtle influence of the hydraulic system characteristics on the closed loop behavior is highly beneficial. Closed loop electrohydraulic control systems were developed in earnest in the 1940s for gun-aiming. But it was not until the advent of the two-stage electrohydraulic servo valve in the 1950s that precise electronic closed loop servo hydraulic control became possible. The diverse range of applications that we see today emerged requiring accurate control of motion at high power, for example in the test and simulation industry, in manufacturing, and in aerospace and defense. The ability to implement sophisticated control laws has increased rapidly since the 1980s with the adoption of digital electronic controllers, allowing control engineers with the appropriate expertise to achieve astonishing levels of performance. For example, megawatts of power can be delivered within milliseconds, positions can be controlled within a few microns, and precise alternating motion at 1000 Hz can be achieved.

In this book, the very wide range of electrohydraulic servo system applications can be appreciated, from those in the materials and structural test industry requiring exceptional precision but small strokes, to materials handling and mobile hydraulic applications where strokes are large but accuracy requirements are less demanding. Also, the reader will be able to ascertain the characteristics of hydraulic components, and how these interact to endow a system with a particular dynamic behavior, which needs to be known or estimated if a model-based controller design approach is to be used. Challenging issues are covered such as coupling within multiaxis systems and also the control of newer energy-efficient architectures, in particular control via variable speed servomotor-driven pumps which avoid the throttling losses in valve-controlled systems. Control methods are described which range from those thoroughly proven in industrial practice to proposed approaches only so far demonstrated in simulation or in laboratory test rigs. This book will be of value both to the novice with an interest in control but new to servo hydraulics and to the seasoned practitioner looking for new perspectives on her or his current control design challenges.

July 2022

Professor Andrew R. Plummer, FIMechE
Head, Department of Mechanical Engineering
Director, Centre for Power Transmission and Motion Control
University of Bath
Bath, UK

Foreword II

The book "Electrohydraulic Servo Systems" authored by Raghavendra DR. a fluid power engineer with vast experience in the domains of fluid power systems and controls, serves as a unique integrated platform. Fluid power systems particularly electrohydraulic control systems are used in a vast range of applications, particularly which demand fast response time and high power levels of the order of several MW.

This book specifically deals with the application of electrohydraulic servo systems, an extremely important part of fluid power. The book aims is to bring together various key aspects of the systems and the layout of the book is well designed with a good blend of theory and applications, which not only describes electrohydraulic servo valve controlled systems, but also provides deep insights into EH servo systems components, their applications in several engineering industries like manufacturing, mobile, motion simulators, material & structure testing, robots, defence aircraft & aerospace. The design aspects and control options for single input and single output (SISO) and multiple input and multiple output (MIMO) systems have been lucidly covered. Fault tolerant methods encompassing model based methods to fault detection and diagnosis are also included. A range of control options in linear/non¬linear regions including the state variable feedback based LQG, model based, sliding mode variable structure and the modern intelligent control methods are dealt with.

EH Servo System is an excellent combination of three core disciplines of power hydraulics, feedback controls systems and transducer technologies and this unique book presents the synergic potential of these systems and highlights a new perspective on the subject. The author has collated the efforts of researchers in the fluid power industry and academies very elegantly and brings out a collection of many interesting and proven solutions with focus on meeting the increasing demands of efficiency and system performance. The challenges of design, synthesis, and analyses to realise an optimum system have been very well addressed. The focus of the book is on the design and testing of electro-hydraulic control systems, which find increasing applications in motion control systems, and also deals with the experimental results to corroborate the designs. The electro- hydraulic servo valves that form the heart of these control systems are sufficiently addressed.

The content of the book is well structured with several practical applications and serves as a Handbook for all practicing engineers involved in the design and development of EH Servo Systems. No doubt that this book can also be treated as a text book on EH Servo Systems for students of mechanical engineering, mechatronics and robotics.

Dr. B. N. Suresh
Indian Institute of Space Science and Technology
Thiruvananthapuram, India

Message from Dr. G. Satheesh Reddy

1. India has undertaken the ambitious program of Atmanirbhar Bharat. One of the pillars of this program is skilled and knowledgeable human capital. Improving the technical skills of our manpower would require effective training on key technological areas like fluid power. Hydraulic servo systems have been traditionally used in high performance applications such as aerospace and defense and also for machine tools, material handling, mobile equipment, mining, and automotive testing. Closed loop hydraulic servo drive technology is finding application in other areas like machine automation and robotics that require greater precision and higher efficiency.
2. Shri. D. R. Raghavenda has decided to bring out a book on servo hydraulics which aims at a contemporary treatment of the subject. He has compiled his experience and exceptional knowledge in the form of a book which aims to demystify the subject for students and fluid power professionals.
3. The author has named the book as *Electrohydraulic Servo Systems* and comprehensively explores the state-of-the-art technology. The work concludes with the

real-world applications of the topics discussed in the book. The book illustrates the subject in a very simple and lucid language which can be easily read and followed by one and all who wish to get an insight into the elements of electro hydraulic servo systems. Nevertheless, he has used his own style of understanding and explaining the things related to this important topic.

4. I whole-heartedly recommend that this book be read and used for further studies and research. I also hope that the book will play its intended part in improving the technical skills of our manpower and help in giving our "Local" products a "Global" branding.

Dr. G. Satheesh Reddy
FNAE, HFCSI, FRIN (London)
FMACANUD (Russia), FAeSI, FRAeS (UK)
HFPMAI, FSSWR, FIET (UK)
FIE, FAPAS, FIETE, AFAIAA (USA)
Secretary, Department of Defence R&D, Chairman, DRDO
Ministry of Defence
New Delhi, India

Preface

Electrohydraulic (EH) servo systems are a blend of oil hydraulics, control systems and transducer technologies. Armed with the power of hydraulics and the control flexibility of electrics, EH servo systems are the preferred drives for many engineering endeavors. The system specifications of efficiency, tracking accuracy, disturbance rejection, and parameter sensitivity for these applications place varying emphasis on each of these requirements. The design of the system is based on the state-of-the-art developments of core technologies. Enormous developments in core technologies have met the increased demands on specifications. The developments are as follows:

- Power elements: pumps servo valves, integrated actuator packages
- Transducer technology, data acquisition, transfer, and processing
- Exponential development in digital data transfer to control action.

Along with these, developments in robotic and aircraft engineering have been adapted to other industrial applications.

These developments and their impact on design for various applications are presented in an array of eight chapters on systems and configurations, EH servo components, applications, design-SISO, control options-SISO, design MIMO, and control options-MIMO. Aspects of fault avoidance, fault detection and diagnosis, and basic needs for fault-tolerant systems form the last chapter.

Bengaluru, India D. R. Raghavendra

Acknowledgements

I am indebted to the following authors for granting gratis permission to use excerpts from their published articles/papers:

- Dr. Andrew R Plummer, Professor and Head of the department of Mechanical Engineering, University of Bath, UK
- Dr. Lasse Schmidt, Associate Professor, Section for Mechatronic Systems, AAU Energy, Aalborg University
- Dr. Tiago Boaventura, Professor, Advanced Robotics Control Lab, Sao Carlos, School of Engineering, University of Sao Paulo
- Dr. Svante Gunnarsson, Professor, Division of Automatic Control, Linkoping University
- Dr. Petter Krus, Professor and Head of Division, Linkoping University
- Dr. Niels Kjostad Poulsen, Professor Emeritus, Department of Applied Mathematics and Computer Science, The Technical University of Denmark (DTU)
- Dr. Srikanth Bashetty, Faculty, Department of Mechanical and Industrial Engineering, A&M University, Texas, USA
- Dr. ir. Maarten Steinbuch, Scientific Director and Distinguished University Professor, Eindhoven University of Technology
- Dr. Shibley Ahmad AL Samaarrie, Professor, Control and Systems Engineering Department, University of Technology, Baghdad, Iraq
- Dr. Edvard Deticek, Faculty of Mechanical Engineering, University of Maribor
- Dr. Jan Swevers, Full Professor, Faculty of Engineering Science, Katholic University, Leuven, Belgium
- Dr. Michele Focchi, Researcher at Universita di Trento and Visiting Scientist at Italiano Tecnologia
- Dr. Joris De Schutter, Full Professor, Mechanical Engineering Division PMA, Katholic University, Leuven, Belgium
- Dr. Wen-Hang Zhu, Department of Electrical and Computer Engineering, British Columbia University, Canada

I wish to thank all the authors of papers/articles who have contributed to the growth of EH servo system technology.

I gratefully acknowledge the support of Sri. P. S. Nair and late Sri. Satish Kanavi of ETA Technology, Bengaluru. I remember the help extended by my colleagues Sri. S. Venkataramiah and Sri. B. K. Anantharaman of CMTI, Bangalore. I express my sincere thanks to Sri. Aravind Srikant of ETA Technology for all the sketches found in the book.

Lastly, I am thankful to Ms. Priya Vyas of my publishers for all the guidance given to me.

Contents

About the Author

D. R. Raghavendra is active in the areas of EH Servo Systems and Robotics on which he has published about 35 papers/articles in national and international forums. He is the recipient of an Appreciation certificate from University of Wisconsin, Madison, US, and an honorary mention award in the Fluid Power Design contest conducted by Hydraulics and Pneumatics Cleveland, US. Two of his papers have won prize paper awards. He has conducted training courses on the subject. He was nominated for the Vasvik Research Award in mechanical engineering branch. He has worked on several national projects involving Servo Hydraulics. After a 12-year stint at Engineering and Design Department of Hindustan Aircraft, Bangalore, as a senior aeronautical engineer, he joined the research and testing department of Central Manufacturing Technology Institute. He retired as Joint Director in the same department. Post retirement, he has served as consultant to Aircraft Accessory and Systems Pvt. Ltd. and is currently an Advisor at ETA Technology Pvt. Ltd. He holds a first-class degree in mechanical engineering from Mysore University.

Acronyms

2ESOBC	Two extended state observer with backstepping controller
ADAS	Advanced driver assist system
ANN	Artificial neural network
ARMA, ARMAX	Autoregressive models
ASC	Automatic strip control
BP	Back passing
BS	British Standard
DCDT	Direct current linear variable differential transducer
DD	Direct driven
DF	Describing function
DOF	Degree of freedom
DRC	Direct robust controller
DTM	Dyadic transfer function matrix
EFB	Electrical feedback
EH	Electrohydraulic
EHSA	Electrohydrostatic Actuator
EHSS	Electrohydraulic servo system
EKF	Extended Kalman filter
ESOBC	Extended state observer with backstepping controller
FBW	Fly-by-wire
FCS	Fire control system
FDD	Fault diagnosis detection
FF	Feedforward
FRF	Frequency response function
GCS	Gun control system
HAA	Hip abduction/adduction
HFE	Hip flexion/extension
IC	Internal combustion
IGBT	Insulated gate bipolar transistor
ILC	Iterative learning control
IM	Independent metering

IMM	Interactive multiple model
IMU	Inertial measuring unit
ISO	International Standard Organization
KFE	Knee flexion/extension
LFW	Linear friction welding
LQG	Linear quadratic Gaussian
LQR	Linear quadratic regulator
LVDT	Linear variable differential transformer
MFB	Mechanical feedback
MIMO	Multiinput multioutput
MISO	Multiinput single output
MPC	Model predictive control
NARC	Nonlinear adaptive robust controller
PI	Proportional integral
PID	Proportional, integral, differential
QFT	Quantitative feedback theory
RBF	Radial basis function
RVDT	Rotary variable differential transformer
SISO	Single input single output
SMC	Sliding mode control
SMISMO	Separate metering in separate metering out
SVD	Single value decoupling
SvSDP	Speed variable switched differential pump
TMD	Tuned mass damper
TVC	Thrust vector control, three variable control
UDEBC	Unknown disturbance estimation-based backstepping controller
UIMM	Updated IMM
UKF	Unscented Kalman filter
UUT	Unit under test
VCD	Voice coil driven
VCP	Virtual cutting point
VDC	Virtual decomposition control

Chapter 1
EH Servo System Configurations

Abstract With a brief introduction to the taxonomy of Servo Systems, servo valve controlled systems, EHSAs and IM valve controlled systems are described. EHSAs have been adapted for applications in Manufacturing, Humanoid Robots, Quadrupeds and prosthetics. IM valve systems address the poor efficiency of Valve controlled systems used in high power Hydraulic Mobile equipment. A novel exhaustive research on use of double pump based EHSA together with IM valves to improve efficiency while maintaining tracking ability of valve controlled systems is cited here.

1.1 Taxonomy of Servo Systems

By definition, servo systems must employ feedback measures and have substantial power amplification. Pneumatic, electric, and hydraulic sources are used in these systems. Electromechanical and electropneumatic systems have limitations in handling power. They lack the quadratic response of hydraulic systems. Hydromechanical systems differ from electrohydraulic (EH) systems in the feedback and command generation methods. Hydromechanical systems that were deployed earlier for hydro copying in machine tools, power steering of automobiles, and in flight controls of aircraft are now outdated.

EH servo systems have higher power-to-weight and power-to size ratios than others. Higher loop gains permitted by these systems yield a faster response and allow closer tracking of command and higher closed loop stiffness. Slow speed running with minimal torque ripple is a desirable feature of these drives. Heat generated due to inefficiency is conveniently cooled through heat exchangers. The nonlinearity of these systems is addressed by modern control options.

EH servo systems are either valve controlled or pump controlled. The valve-controlled systems close-couple the valve/actuator, have a faster response, and work at constant supply pressure. This renders them inefficient, thereby requiring larger coolers. Pump-controlled systems have poorer response because of large trapped volumes and are efficient as the drive power is closely matched to drive requirements.

D. R. Raghavendra, *Electrohydraulic Servo Systems*, Springer Tracts in Mechanical Engineering, https://doi.org/10.1007/978-981-19-8065-7_1

Additionally, pump-controlled systems need an extra charge pump, servo valve, and actuator for stroking the pump displacement mechanism.

Electrical servo motors driving a hydraulic pump that feeds an actuator directly without a servo valve are designated Electrohydrostatic Actuator (EHSA) systems. These drives address the inefficiency of servo valve-controlled systems and claim to match the efficiency of electromechanical systems with added hydraulic benefits. EHSAs are also proposed for higher-power applications where the size of electric motors is acceptable.

The inefficiency of valve-controlled systems has been addressed by recent independent metering (IM) valves and proposed for power-intensive mobile applications. IM valves are also referred to as separate metering in separate metering out (SMISMO) valves and programmable valve systems by some researchers (Fig 1.1).

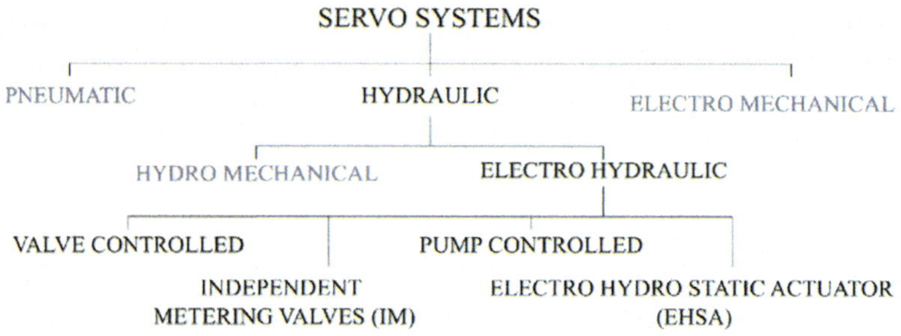

Fig. 1.1 Taxonomy of servo systems

1.2 Valve Control

The general configuration of a valve-controlled EH servo system is shown in Fig. 1.2. Servo valve, the power amplifier feeds the actuator driving the load based on some function of command and feedback signals. Feedback transducers measure the controlled parameters, which can be linear position or force, rotary position, and speed or torque or pressure output of the actuator. The design of the system entails the sizing of all elements, including the hydraulic power supply, to meet the control of the load parameter.

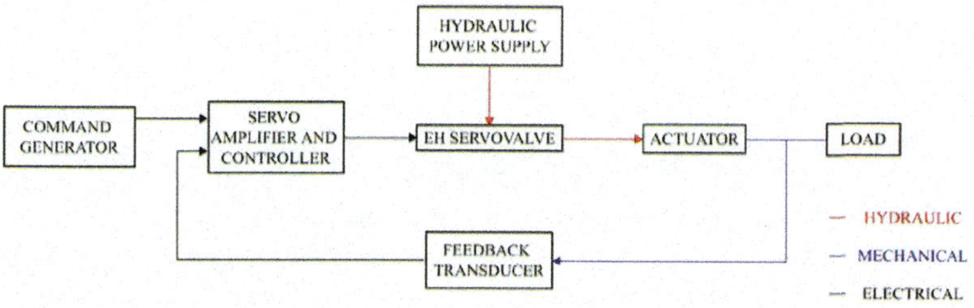

Fig. 1.2 Elements of EH servo systems

1.2.1 *Linear Position*

The position transducer measures either the actuator or load position. Load position feedback leads to more accuracy, but the backlash of mechanical elements may cause instability. Actuator position feedback avoids these stability issues at the expense of accuracy. The position transducer is selected based on the desired accuracy level of application, a minimum of one grade higher accuracy. When the transducer is digital, the system includes an analog to digital converter, as the servo valve needs an analog input. The power elements are sized to meet the *load locus* (force/velocity requirement over range of working) (Figs. 1.3 and 1.4)

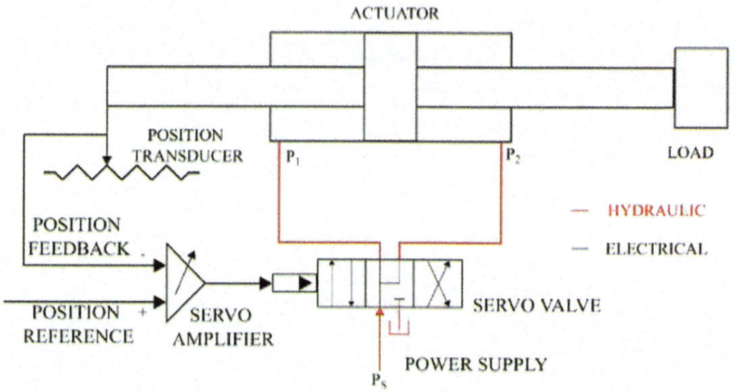

Fig. 1.3 Linear position control

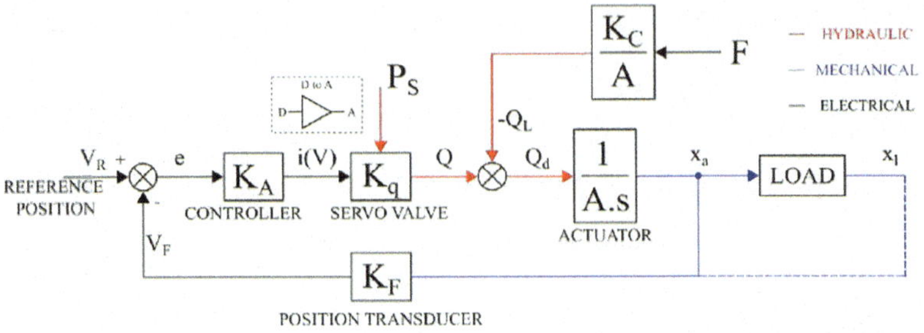

Fig. 1.4 Block diagram of linear position control

Tracking accuracy, resistance to load disturbance, and sensitivity to parameter variations are critically dependent on the loop gain of the system. The system gain needed is limited, generally by the poorly damped load/actuator resonance. Gain can be substantially increased by judiciously adding acceleration and velocity feedback to the system.

1.2.2 Force Control

Material and structure testing for strength, fatigue, or modal analyses requires force control (Figs. 1.5 and 1.6).

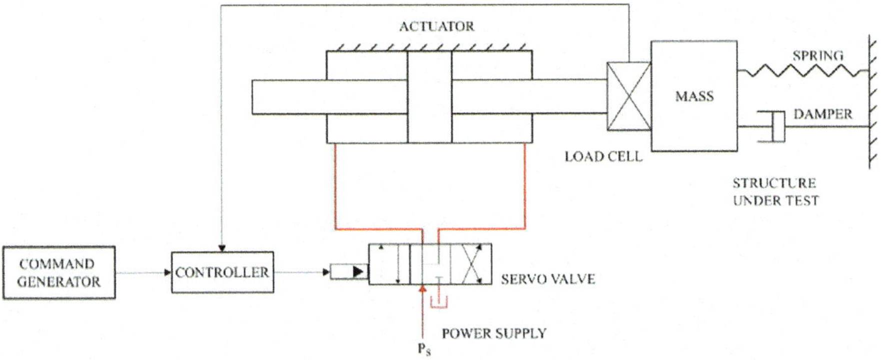

Fig. 1.5 Generic force control system

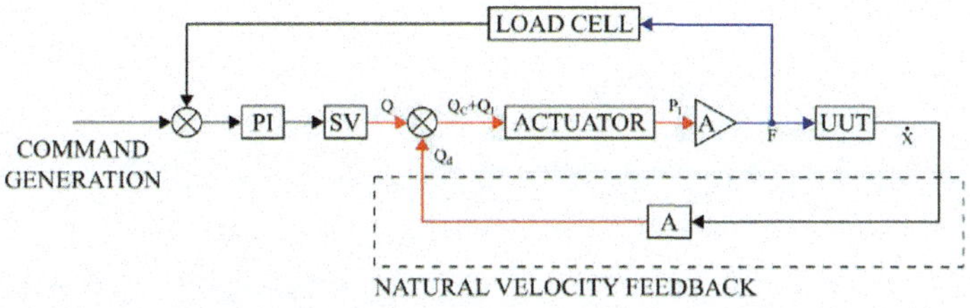

Fig. 1.6 Block diagram of the force control system

Velocity in force control acts as a disturbance causing errors in control. Force control systems are subjected to natural velocity feedback. The controllers are designed to compensate and feed to the valve an additional input for displacement flow demanded by the system [1].

Load cells, which are strain gauge based, meet the requirements of most applications. Since most systems are computer controlled for preloop processing and posttest analysis purposes, both analog-to-digital and digital-to-analog converters form part of the system. Piezo-based force transducers are used for quasi-static and dynamic applications.

1.2.3 Pressure Control

Structural integrity of all the hydraulic components is tested for proof and burst ratings as per standards. Some elements like hoses, fittings, and actuators are also tested for their impulse pressure rating. These tests call for pressures higher than the system pressure. The higher pressures are derived from air/hydraulic or hydraulic/hydraulic intensifiers (Figs. 1.7 and 1.8).

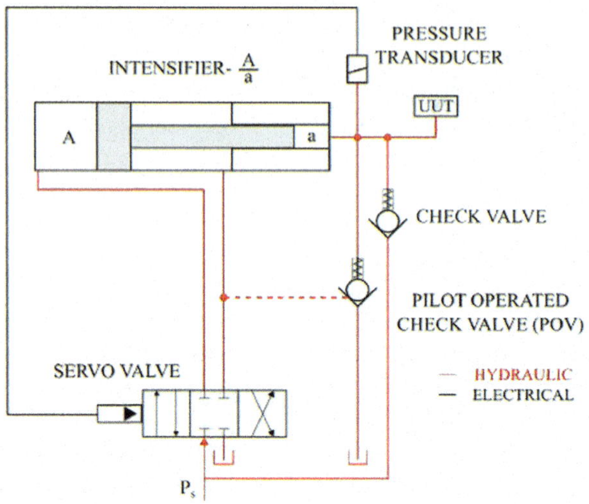

Fig. 1.7 Pressure control with intensifier

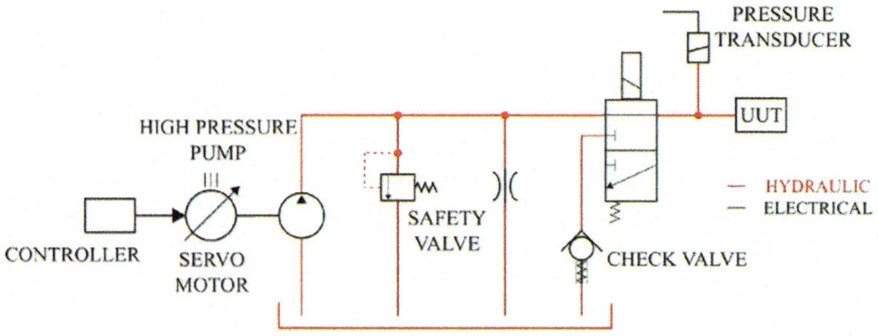

Fig. 1.8 Pressure control with high-pressure pump

High-pressure pumps, such as check ball pumps working with a pressure transducer in a closed loop configuration, can meet the requirements of pressure control.

1.3 Hydrostatic Drives

Rotary speed drives use servo-controlled pump and motor combination with either a tacho generator or speed derived from a high resolution encoder for speed control. The control loop uses an integrating amplifier facilitating control.

The servo pump is coupled to a charge pump that delivers pressurized oil to a servo valve. The servo valve controls the swash plate angle of an axial piston pump or the eccentricity of a radial piston pump. Displacement control is offered with pressure compensation, flow, or constant power options.

The servo pump is connected to a fixed or variable displacement motor in either open-circuit or closed-circuit configurations. In an open circuit, the pump draws oil from a reservoir, and the motor dumps back the oil into the reservoir. These work at a maximum pressure of 210 bars, use a large reservoir, and easily maintain the oil temperature (Fig. 1.9).

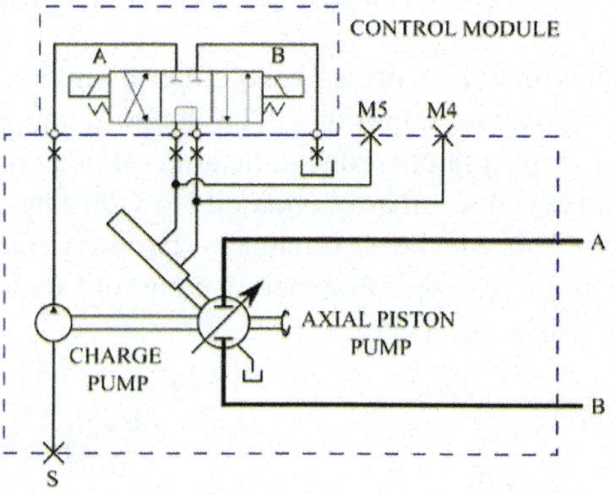

Fig. 1.9 Servo-controlled axial piston pump

Hydrostatic closed-circuit drives work at higher pressures, up to 350 bar, have smaller reservoirs, and are expensive. The pump and motor are connected by long tubes or hoses and, in a few cases, are closely coupled.

The circuit includes safety relief valves, counterbalance valves for load holding, and hydraulically operated brakes to cater to any inadvertent hose failure. The circuit includes a small heat exchanger for temperature control (Fig 1.10).

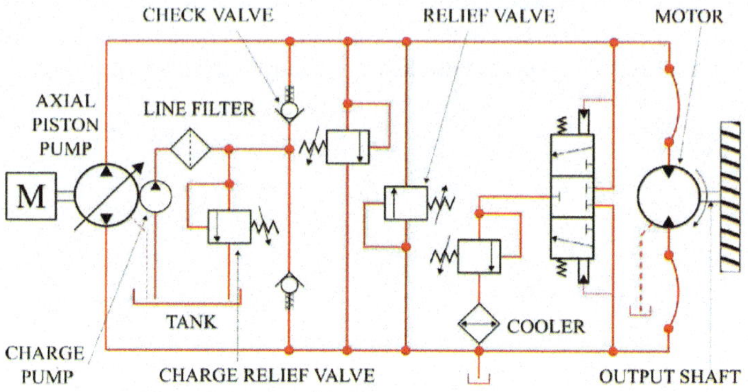

Fig. 1.10 Hydrostatic closed-circuit drive schematic

1.4 EHSA Drives

The concept of an integrated electrical servo motor and a hydraulic pump driving an actuator directly without the use of a servo valve was applied for Airbus Flap actuation. Studies for industrial applications were made with variants of pump, motor, and circuit combinations with different degrees of integration. EHSA is a compact drive with comparable efficiency as electromechanical drives with added advantages of robustness, overload protection, and higher force capability of hydraulic drives [2].

The design variants of EHSA drives have either a single pump or dual pumps in open- or closed-circuit configurations. The single pump with inverted shuttle valves (Fig. 1.11) is compact but is prone to heating oil in heavy duty cycle applications. Suppliers specify the use of ***isoviscous*** oils to withstand higher temperatures. Open circuits overcome this with heat exchangers. The drives facilitate energy-saving regenerative circuits to be realized. Regenerative circuits are used with unbalanced actuators to increase extension speeds.

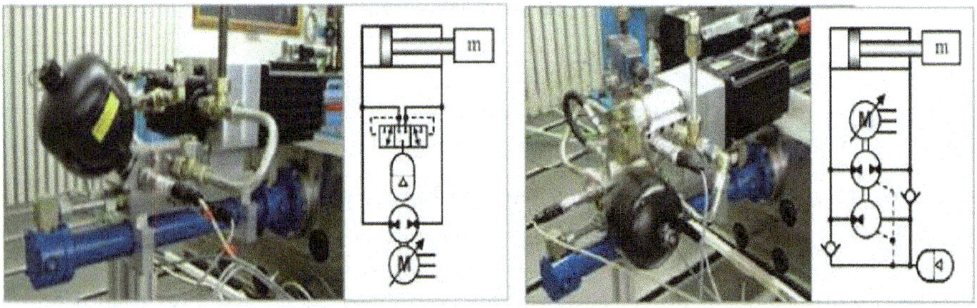

Fig. 1.11 Inverse shuttle valve EHSA [2] double pump [2]

A study made at ETA Technologies (Fig. 1.12) for application to low tonnage rotary friction welding revealed the feasibility of switching control modes from position to force and vice versa with desired accuracy levels [3].

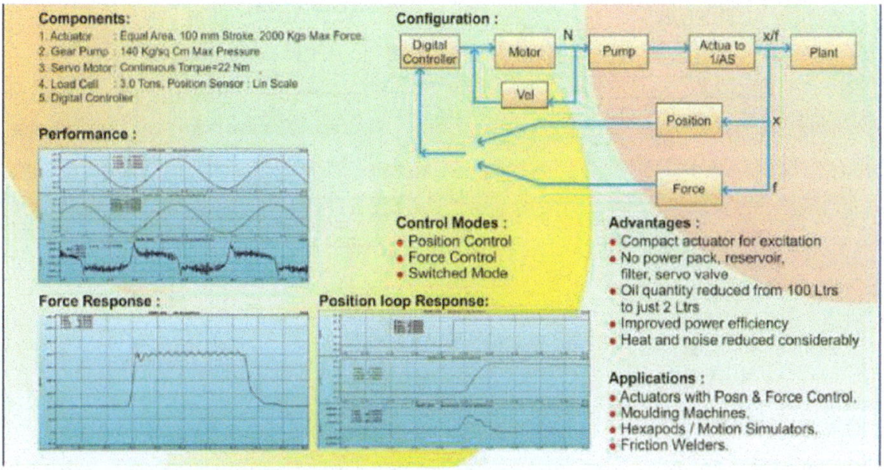

Fig. 1.12 EHSA study with inverse shuttle valves [3]

Several leading hydraulic component manufacturers have added these drives to their product range.

Electric servo motors driving fixed delivery gear pumps or variable delivery piston pumps [4] for higher-power levels provide efficient hydraulic power sources for test rig applications where the size of the motor may not matter. They enable efficient ways to meet the varying flow and pressure requirements of these applications.

1.5 IM Valves

Energy-intensive mobile hydraulic systems place high premium on the efficiency of the systems. The servo valve-controlled systems operating at a constant power supply dissipate energy when working under nonideal working conditions of load and velocity. IM valves controlling individual actuator chamber pressures optimally vary the supply pressure to meet load conditions and thus open up a number of energy-saving possibilities.

These electrically modulated valves are available in two port/two position (cartridge type), four port/three position, and three port/three position variants. At the actuator level, the drivers are capable of working in four quadrants, facilitating regenerative and energy recuperation with circuitry (Figs. 1.13, 1.14 and 1.15).

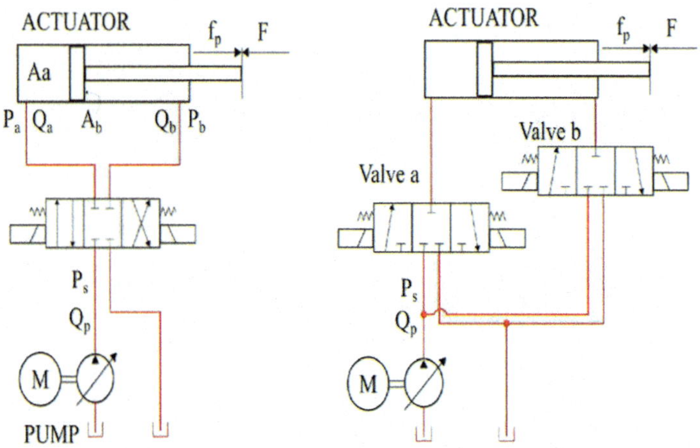

Fig. 1.13 Conventional and IM/SMISMO drive configurations **a** Conventional drive. **b** IM drive with three-position/three-port valves

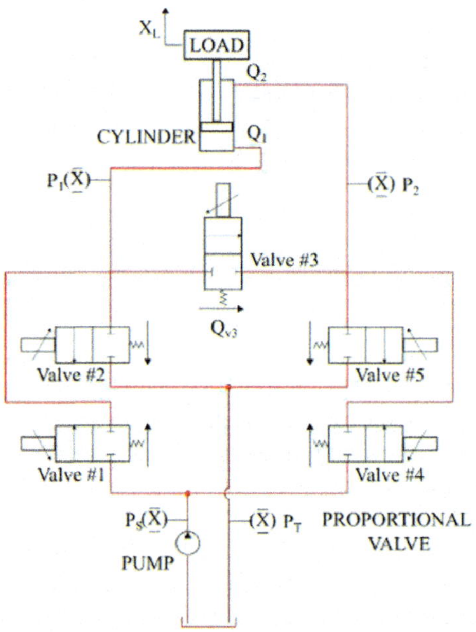

Fig. 1.14 Programmable valves in two-position/two-port configuration

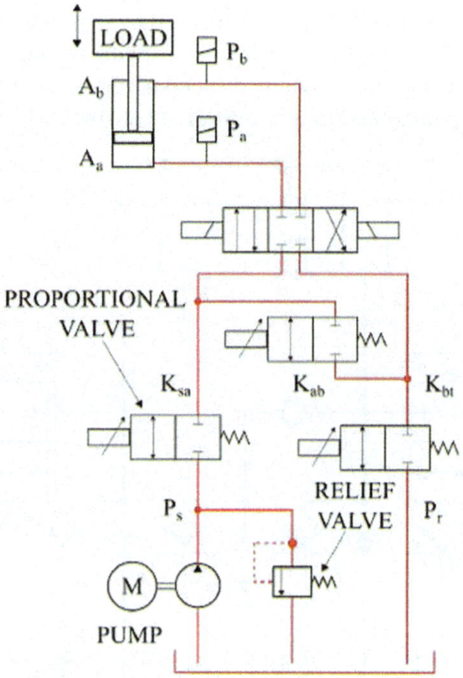

Fig. 1.15 Programmable valves with 1-direction control valve [5]

Nguyen et al. [5] cites the above advantages of IM valve technology and notes the demerits as follows:

- High cost of proportional valves and their number required
- MIMO control complexity of individual valves
- Switching mode causing variations in velocity.

The authors describe circuits to reduce the number of proportional valves from five to one. These circuits use a combination of proportional and normal solenoid-operated directional control valves to achieve control objectives.

The two-level control of IM valves permits selection of the mode of operation and individual chamber pressure control.

Koivumaki et al. [6] presents the methodology of control of individual valves and smooth switching over. Focused research is pursued on this subject at Universities and Research Centers in Finland, Sweden, Germany, the USA, China, and Korea.

1.6 EHSA with IM Valves

EHSA and IM valve-controlled drives work at lower-pressure levels to save energy. This renders them susceptible to cavitation and poor tracking ability because of the reduced bulk modulus of oil. These are addressed by [7], suggesting a novel concept of a dual pump EHSA combined with IM valves with an additional pump and with anti-cavitation valves. The drive is named speed variable switched differential pump (SvSDP). The third pump ensures a minimum oil stiffness (Fig 1.16).

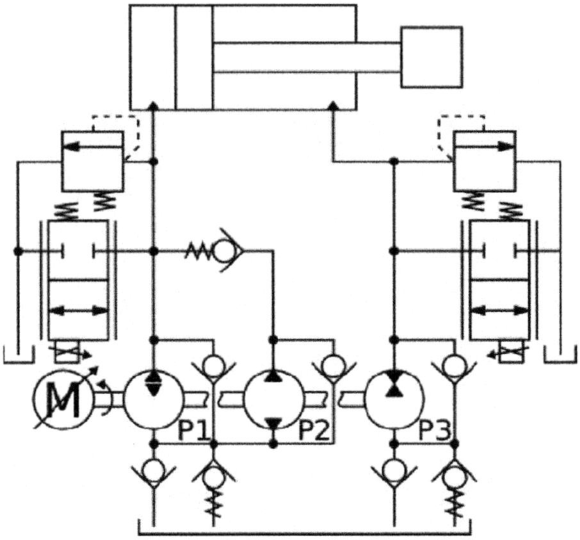

Fig. 1.16 SvSDP concept [7]

The exhaustive study details the modeling of electric, hydraulic, and mechanical elements and the simulation and verification of the models and control strategy. The controller accounts for MIMO control of IM valves and includes the additional decoupler for the loading system. The study concludes the matching tracking performance of SvSDP with that of valve-controlled systems. It notes the loss of quiescent power loss under static conditions and the additional expense of drive.

Schmidt et al. [8] focuses on the double pump EHSA combined with IM valve configuration, noting the following:

- Near-zero quiescent power loss under static conditions due to IM valves
- Four-quadrant operational capability of drive and
- The common return path to the reservoir permits oil cleanliness and temperature control (Fig. 1.17)

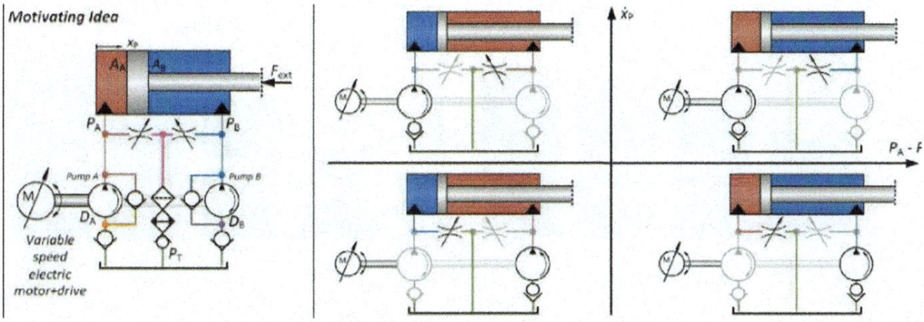

Fig. 1.17 Double pump EHSA with IM valves. Self-locking, four-quadrant operation [8]

Based on the self-locking nature, the study proposes six alternative configurations to include energy regeneration and recuperation options for the drive (Fig. 1.18)

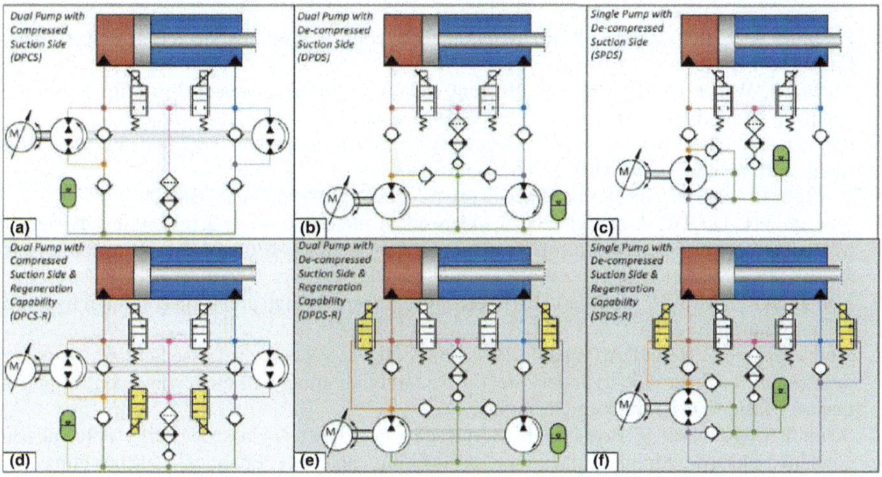

Fig. 1.18 Configurations. D, E, F with regeneration capability [8]

The study presents a loss analysis and energy recovery potential of the drives with comparison to valve-controlled drives. The drivers provide an optimizing option of tracking performance and energy savings.

The study presents the modeling and control of the drivers (DPCS-R, DPDS-R), enabling tracking performance and energy-saving potential. The study claims the drives provide proper motion performance, lower-pressure level control, and superior energy efficiency in the majority of the operating range when compared to valve-controlled drives.

The economic viability of the DPCS-R and DPDS-R drives considering the cost of all elements and noting that the drive needs one servo motor per axis are indicated in Fig. 1.19.

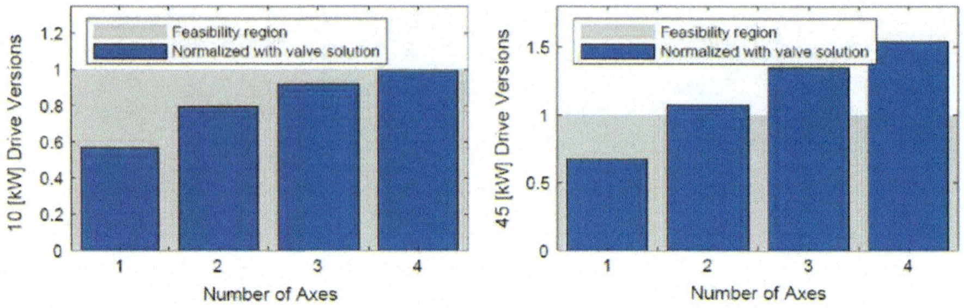

Fig. 1.19 Economic viability of drives [8]

References

1. Plummer AR (2007) Robust electro hydraulic force control. Proc Inst Mech Eng. J Syst Control Eng 221:717
2. Michel S, Weber J (2012) Energy efficient electro hydraulic compact drives for low power applications. In: Fluid power motion and control, Bath, UK
3. Raghavendra DR, Kanavi S (2016) EH servo systems—developments. In: Fluid power technical seminar, Bangalore, June 2016
4. RexRoth Bosch Sytronix. Variable speed drives catalog. R999000332.2016-09
5. Nguyen TH, Do TC, Ann KH (2020) Independent metering valve: a review of advances in hydraulic machinery drive control. Soc Hydraul Constr Mach 17(4):54–71. https://doi.org/10.7839/KSFCA.2020.17.4.054
6. Koivumaki J, Zhu WH, Mattilla J (2019) energy efficient and high precision control of hydraulic robots. Control Eng Pract
7. Sloth-Osgaard P, Hertz RA, Valentin-Pederson S (2017) Supervisor: Lasse Schmidt. Topology optimisation of hydraulic cylinder direct drive targeting energy efficiency and control performance. Master thesis, Aalborg University
8. Schmidt L, Ketelsen S, Brask MH, Mortensen KA (2019) A class of energy efficient self-contained electro-hydraulic drives with self-locking capability. Energies 12:1866. https://doi.org/10.3390/en12101866.MDPI

Chapter 2
EH Servo System Components

Abstract The range of Hydraulic components, Feedback Transducers and Controllers are tabled with their performance characteristics. A distinction is made between the Torque Motor driven and Voice Coil Driven servo valves.IM valves, Integrated Servo Drive packages, Multi- axis IMUs are listed.

2.1 Electrohydraulic Servo Systems (EHSS)

An EHSS is an ensemble of hydraulic power elements, feedback elements, and controllers. Different configurations call for different elements. To synthesize an EHSS demands knowledge of the design variants of the components as well as their performance characteristics.

2.2 Servo Valves and Servo Proportional Valves

By responding to electrical signals, servo valves serve as power amplifiers and form the heart of these systems. The hydraulic part of the valve is a spool moving inside a metered bush. Precision machined-controlled edges yield critically lapped, underlapped, and overlapped valves and enable functional requirements of four quadrant operation as well as meet regeneration for energy savings.

A basic distinction is made between the original torque motor-driven servo valves and the developed voice coil-driven (VCD) ones, also referred to as direct-driven (DD) valves. While the torque motor-driven servo valves depend on designs of hydraulic amplifiers for good response, VCD (DD) valves have higher spool stroking forces. Torque motor-driven valves are rated at a 70 bar valve pressure drop. The VCD valves are rated at 70 bar for flows up to 60 lpm and at a 10 bar valve pressure drop for higher flows. These servo proportional valves work at 350 bar supply pressure and use axial force compensating grooves on spools in the design. They limit the maximum flow to 3 times the rated one for stable operation.

D. R. Raghavendra, *Electrohydraulic Servo Systems*, Springer Tracts in Mechanical Engineering, https://doi.org/10.1007/978-981-19-8065-7_2

2.2.1 *Torque Motor-Driven Servo Valves*

The original wet torque motors have since been replaced by dry ones.

The two-stage dry torque motor-driven servo valves are classified based on the hydraulic amplifier and the spool position feedback design (Fig. 2.1).

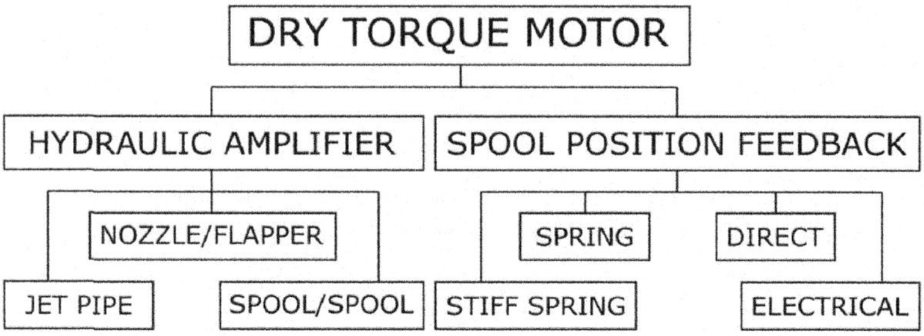

Fig. 2.1 Design variants of torque motor-driven valves

The double nozzle flapper design using fixed restrictors was subjected to hard over failure, which is not acceptable for certain applications. This is overcome by fluid filtration or by single entry jet pipe servo valves. The spool and spool design suffers from accumulated silt level contamination affecting threshold performance. This is countered by use of a high-frequency, low-amplitude input Dither signal to the valve (Figs. 2.2, 2.3, and 2.4).

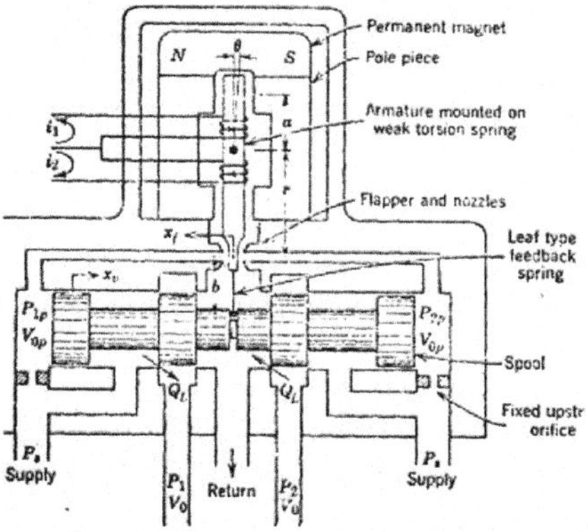

Fig. 2.2 Nozzle/flapper [1]

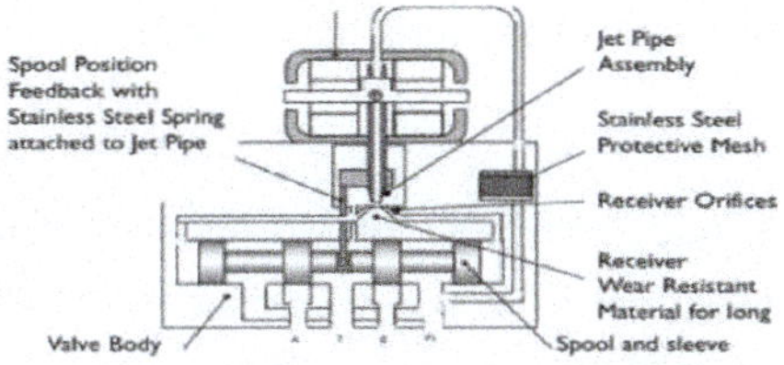

Fig. 2.3 Jet pipe valve [2]

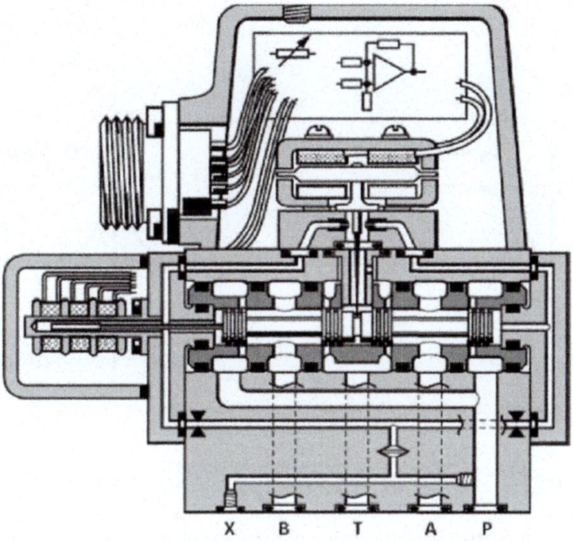

Fig. 2.4 Nozzle flapper valve with electrical spool position feedback [3]

Electrical spool position feedback with an LVDT facilitates control filter design to improve valve response by canceling the one-stage resonance limiting gain.

2.2.2 *Performance*

The servo valve static performance is defined by

$$Q_l = k \times i \times \overline{P_s - P_l}$$

where k is the sizing constant, i is the input current, P_l is the load pressure, and P_s is the supply pressure.

Servo valves are rated for flow Q_r at a valve pressure drop of 70 bars for rated 100% input.

Figure 2.5 defines the static characteristics of valve.

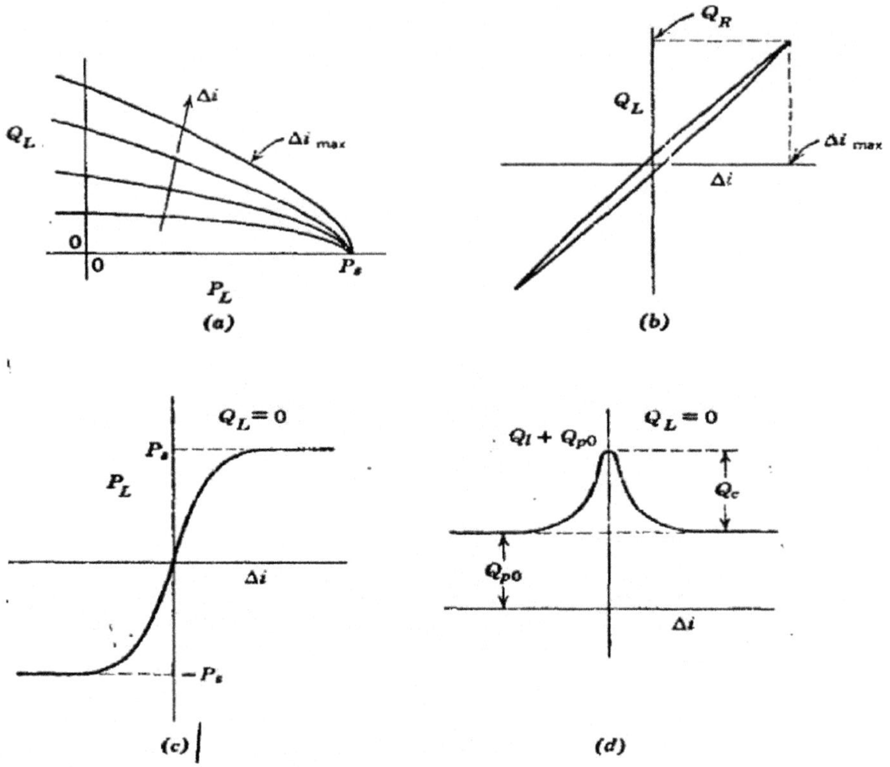

Fig. 2.5 Static characteristics of servo valve [1]. **a** P_{l} versus Q_{l}, **b** input versus Q_{l} at $P_{\mathrm{l}} = 0$, **c** input versus Pl at $Ql = 0$, **d** leakage flow versus input at $Q_{\mathrm{l}} = 0$

A plot of load flow Q_{l} versus load pressure P_{l} for different inputs is shown in Figure 2.5a.

A plot of Q_{l} versus i input at P_{s} at no load ($P_{\mathrm{l}} = 0$) is the normal flow gain and a plot of P_{l} versus i at $Q_{\mathrm{l}} = 0$, its pressure gain is referred to as K_q while is K_p its pressure gain.

The quoted value for K_p is 30–80% of P_{s} for 1% valve input. This high-pressure gain is the reason for the high acceleration capability of hydraulic servo systems. Kce is the flow pressure coefficient defined.

These coefficients are necessary for the derivation of white box models of the system.

Another important characteristic of valves is null leakage, which should be low to keep the quiescent power loss minimum (Fig. 2.5d).

The servo valve dynamic performance is indicated, in catalogs, in terms of frequency or step response in time. Figure 2.6 shows typical plots. The dynamic performance is a function of the supply pressure to the pilot stage of the valve and input amplitude. It is modeled as a damped second-order quadratic.

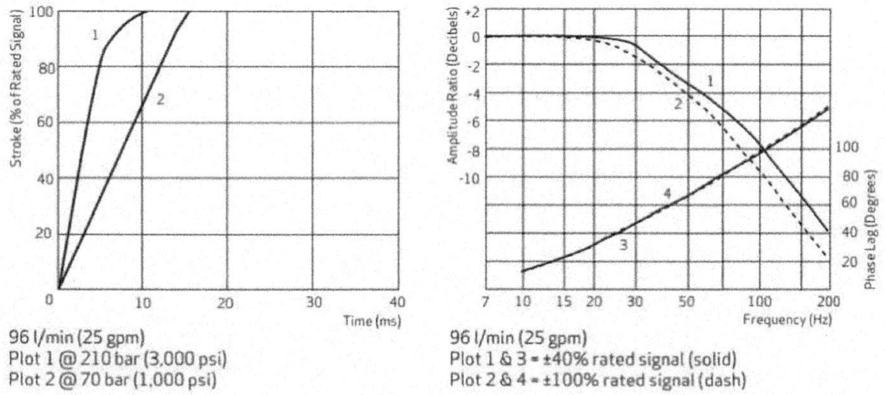

Fig. 2.6 Dynamic characteristics of the servo valve [4]

A comparison of the performance of these valves is given in Table 2.1. Two-stage servo valves working at 210 bars are available from 4 to 63 lpm rated flows and from 96 to 230 lpm in high flow versions.

Table 2.1 Comparison of performance of valves [5]

Valve type	Direct drive valve (DDV)		Two-stage servovalve		
	Open loop proportional Valve	Position controlled proportional valve	Force motor DDV	Hydraulic pilot, mechanical feedback (MFB)	Hydraulic pilot, electrical feedback (EFB)
Spool actuation	Proportional solenoid, open-loop	Proportional solenoid, dosed-loop	Linear force motor (voice coil)	Hydraulic, mechanical feedback	Hydraulic, electrical feedback
Actuation force (N)	< 50	~50	~20	~500	~500
Static accuracy					
Hysteresis (%)	5+	2	0.2	2	0.2
Dynamic response					
Step response (100%) (ms)	100	50	15	10	3
90° phase lag frequency (Hz)	5	10	50	100	200
Cost	Very low	Low	Medium	High	Very high
Size	Large	Very large	Very large	Small	Medium

2.2.3 *Efficiency*

The efficiency of servo valve-controlled systems is poor when fed with constant displacement pumps even at 100% valve input (Fig. 2.7).

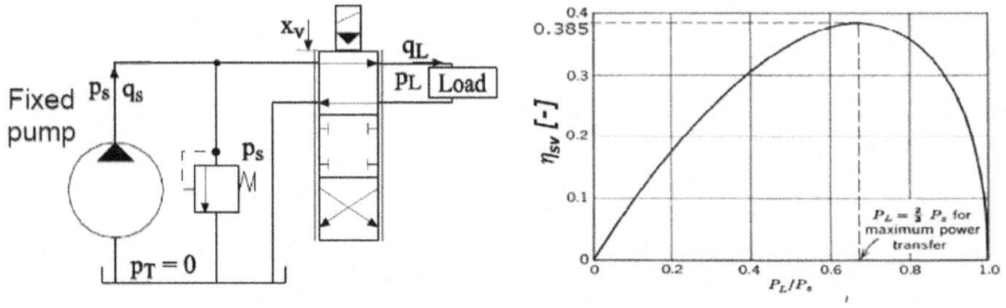

Fig. 2.7 Efficiency versus Pl/P_S [1]

Variable displacement pressure compensated pumps are preferred as a power source as they improve efficiency. Inefficient power heats up the fluid medium and calls for higher cooling capacities.

2.2.4 *3-Stage Servo Valves*

For applications demanding higher flow rates, three-stage valves are designed. A two-stage pilot valve drives a metered, sized spool controlling its position in a closed loop. This loop is tuned in the servo amplifier to maximize valve response. Three-stage valves working at 210/280 bar supply pressures are available in flow ranges from 150 to 3000 lpm (Fig. 2.8).

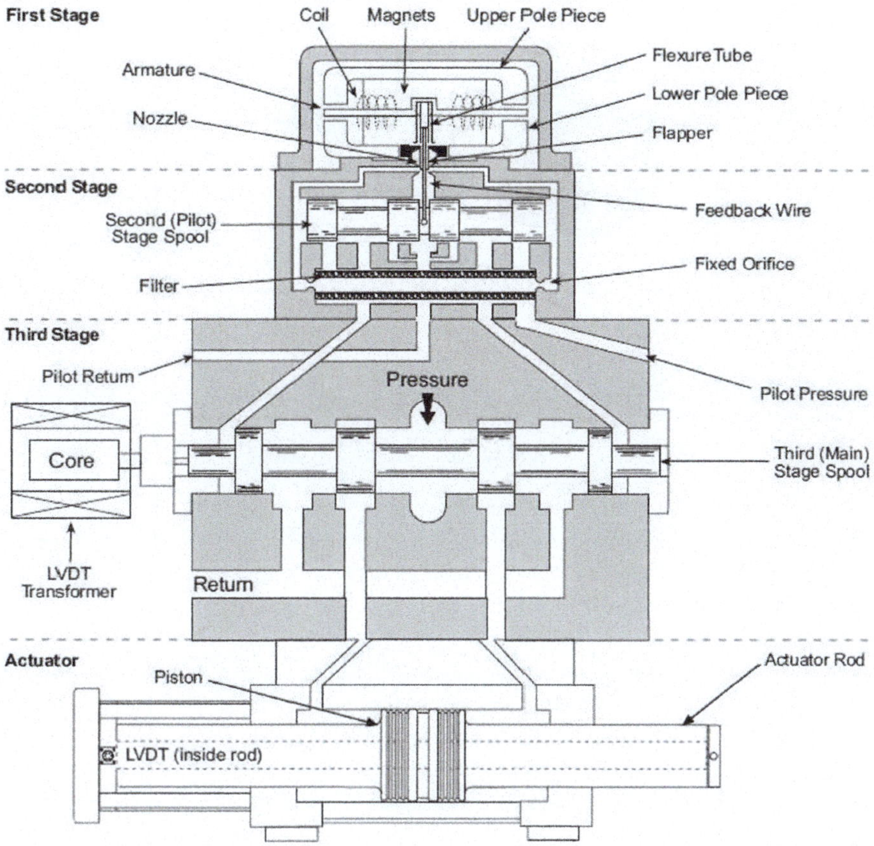

Fig. 2.8 Three-stage servo valve [6]

2.3 Servo Proportional Valves (VCD)

The development of linear voice coil drivers leads to a series of servo proportional valves combining the response of servo valves and a low, 10 bar pressure drop across metering edges of proportional valves. These valves claim superior static and dynamic characteristics. Vesicles that are relatively large and heavy are directly operated or pilot operated for higher flows. Directly operated valves are rated at a 70 bar valve pressure drop, while pilot operated valves are rated at 10 bar. One directly operated VCD valve is rated 80 lpm at a 10 bar pressure drop with the maximum flow limited to 250 lpm. The input power from the amplifier to the valve is high (Figs. 2.9 and 2.10; Table 2.2)

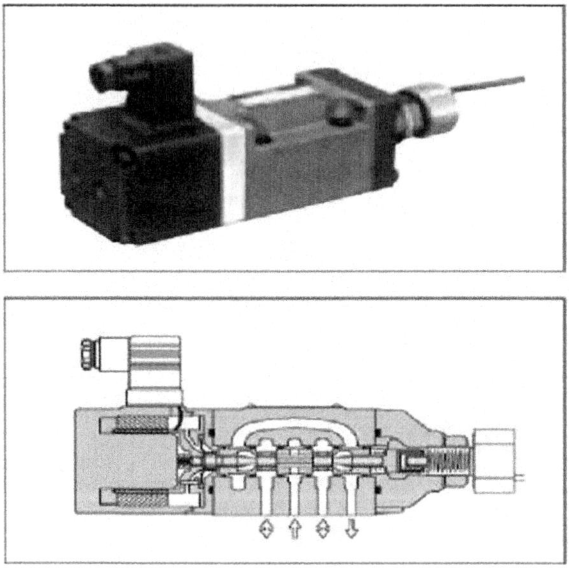

Fig. 2.9 Directly operated VCD valve [7]

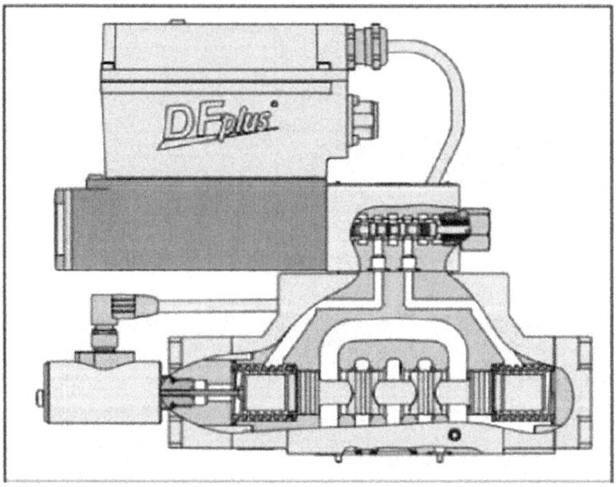

Fig. 2.10 Pilot operated VCD valve [8]

Table 2.2 Performance of direct operated VCD valve [7]

Description		Model numbers	
		LSVG-03-4/10/20/40	LSVG-03–60
Rated Flow @ ΔP = 7 MPa (1020 PSI)[a]		4, 10, 20, 40 L/min (1.06, 2.64, 5.28, 10.57 U.S.GPM)	60 L/min (15.85 U.S.GPM)
Max. operating pressure		35 MPa (5080 PSI)	
Proof pres. at return port		35 MPa (5080 PSI)	
Drain port (Y) permissible back pres.[b]		0.05 MPa (7 PSI)	
Null leakage (@Ps = 14 MPa (2030 PSI) 32 in mm^2/s (150 SSU)		1.7 L/min (0.45 U.S. GPM) or less	
Hysteresis		0.1% or less	
Step Response (00 $\Longleftrightarrow$ 100%, typical)[c]		2 ms {3 ms}	3 ms {4 ms}
Frequency response (± 25% amplitude, typical)	Gain: −3 dB	350 Hz {300 Hz}	330 Hz {240 Hz}
	Phase: −90°	450 Hz {370 Hz}	410 Hz {330 Hz}
Vibration proof[d]		Frequency: 10–60 Hz. Amplitude: 4 mm (0.157 in.). Acceleration 7.8–282 m/s^2 (25.6–925 $ft./s^2$) Frequency: 61–2000 Hz. Amplitude: 4–0.0038 nun (0.157–0.00015 in. (Acceleration: 294 m/s^2: (965 $ft./s^2$)	
Protection		IP 64	
Ambient temperature		−15 to + 60 °C (5 to 140 °F)	
Spool type		Neutral/zero lap	
Spool stroke to stops		± 0.5 mm (± 0.0197 in.)	+ 7.5 mm (+ 0.0295 in.)
Linear motor specification	Current	2 A [Max. 6 A]	
	Coil resistance	4.5 Ω [at 20 °C (68 °F)]	
Mass		5 kg (11.0 lbs.)	
Applicable servo amplifier		AMLS-A-D*-*-10	AMLS-B-D*-*-10

These valves provide spool position data monitoring for diagnostic purposes and an option of the fourth position in case of power failure. The valves function in the normal four quadrants and are an option to work in regenerative mode (Table 2.3).

Table 2.3 Performance of pilot operated VCD valves [8]

Hydraulic		NG10 (CETOP 05)	NG16 (CETOP 07)	NG25 (CETOP 08)	NG32 (CETOP 10)
Maximum operating pressure		Internal pilot drain P. A. B, X: 350 bar (5075 PSI). T. Y: 35 bar (500 PSI)			
		External pilot drain P. A. B. T, X: 350 bar (5075 PSI); Y 35 bar (500 PSI)			
Fluid		Hydraulic oil as per DIN 51524 … 51535, other on request			
Fluid temperature		−20 °C to + 60 °C (−4 °F to +140 °F)			
Viscosity permitted		20 to 380 cSt (mm^2/s) (94 to 1727 SSU)			
Viscosity recommended		30 to 80 cSt (mm^2/s) (140 to 375 SSU)			
Filtration		ISO Class 4406 (1999) 18/16/13 (acc NAS 1638: 7)			
Nominal flow at $\Delta p = 5$ bar (73 PSI)	Per control edge[a]	120 LPM (32 GPM)	200 LPM (53 GPM)	400 LPM (106 GPM)	1000 LPM (265 GPM)
Maximum recommended flow		250 LPM (66 GPM)	600 LPM (159 GPM)	1000 LPM (265 GPM)	3000 LPM (794 GPM)
Leakage at 100 Bar (1450 PSI)	Overlapped spool	0.2 LPM (0.05 GPM)	0.2 LPM (0.05 GPM)	0.6 LPM (0.16 GPM)	1 LPM (0.26 GPM)
	Zerolapped spool	0.9 LPM (0.24 GPM)	0.9 LPM (0.24 GPM)	1 LPM < 0.26 GPM)	—
	Pilot	< 1 LPM (0.26 GPM)			
Pilot supply pressure		20 Bar (290 PSI) to 350 Bar (5075 PSI)			
Pilot flow, step response at 210 bar (3045 PSI)		10 LPM (2 6 GPM)	12 LPM (3.2 GPM)	24 LPM (6.3 GPM)	40 LPM (10.6 GPM)
Static/dynamic					
Step response at 100% stroke		10 ms	13 ms	19 ms	45 ms
Frequency response ± 5% at 210 bar (3045 PSI)	Amplitude	128 Hz	95 Hz	95 Hz	40 Hz
	Phase	118 Hz	95 Hz	90 Hz	75 Hz

2.4 Piezo-Based Servo Valves

Research is on to develop piezo-based servo valves with superior dynamic performance. Several design variants using piezo actuators to stroke the first-stage flapper or the main stage spool are being tested. Reference [9] reviewing the research on the subject notes the different piezo actuators under trial (Fig. 2.11 and Table 2.4).

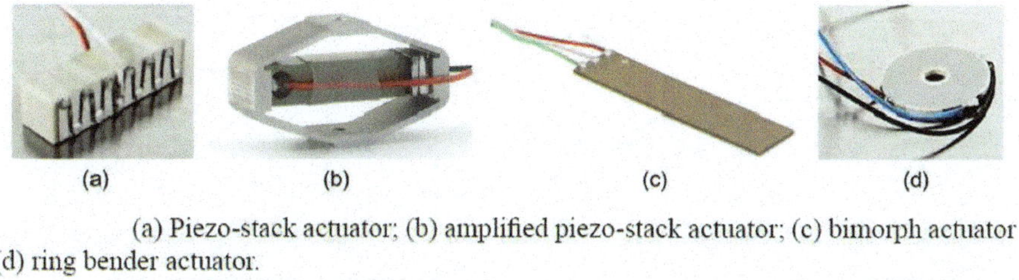

Fig. 2.11 Piezo actuators for servo valves [9]

Table 2.4 Comparison of piezo actuators [9]

Type of piezoelectric actuator	Advantages	Disadvantages
Piezo-stacks	Very high actuation forces	Very low ratio of displacement to size, high hysteresis
Amplified piezo-stacks	High actuation forces, medium displacement	High complexity, low ratio of displacement to size, high hysteresis
Rectangular benders	High displacement	Very low actuation forces, high hysteresis
Ring bender	Medium displacement, medium actuation forces	Lower forces than piezo-stacks and lower displacement than rectangular benders, high hysteresis

The use of ring benders to control the first-stage flapper is detailed. A piezo actuator stroking the metering bush of the valve while the hydraulic amplifier moves the main stage spool, referred to as hybrid design, is described in [10]. The use of two independent transducers or one measuring relative movement is considered (Figs. 2.12 and 2.13).

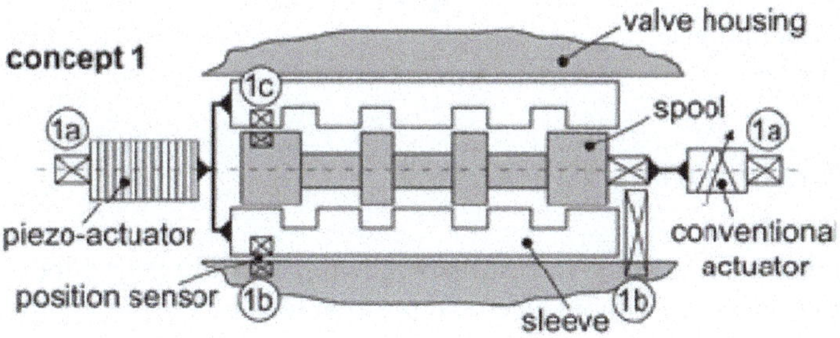

Fig. 2.12 Piezo stack driving metering bush in hybrid design [10]

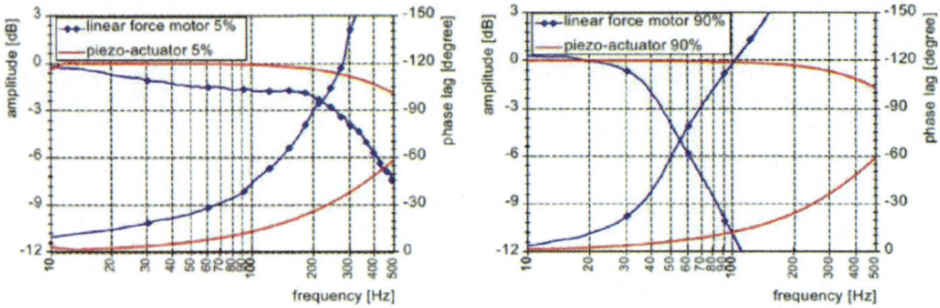

Fig. 2.13 Performance comparison [10]

2.5 Proportional Valves

The development of proportional solenoids led to a series of proportional valves capable of working in open or closed loop configurations. The valves are overlapped, but the servo amplifier design permits control in a closed loop. These valves for directional flow or pressure were developed to enable control of high-power hydraulics with electrical signals.

2.5.1 Proportional Direction Control

These four port/three position valves combine direction and flow control functions. They are available in direct or pilot operated versions for a wide range of flow rates. The spools are either spring centered or controlled in a closed loop with LVDT feedback. The latter claim a faster response. These are available in various mid position spool options.

2.5.2 Pressure Control

Pressure relief and reducing valves, likewise, are either direct or pilot operated for higher flow rates, with/without electrical position feedback of the control element.

2.5.3 *Flow Control Valves*

These two port/two position valves control flow proportionally to the input electrical current and are available with pressure compensators.

2.6 IM Valves

IM (SMISMO) valves are classified as 3/3 or 2/2 valves and have spools and poppets. Spool valves provide flow accuracy and are stable but are sensitive to contamination and subject to internal leaks. Poppet valves have lower flow accuracy but are not subject to leakages.

Slip in cartridge poppet valves that are widely used in IM configuration is available in a wide range of flow ratings at pressures up to 280 bars controlled by proportional solenoids or VCD driven [11] (Fig. 2.14).

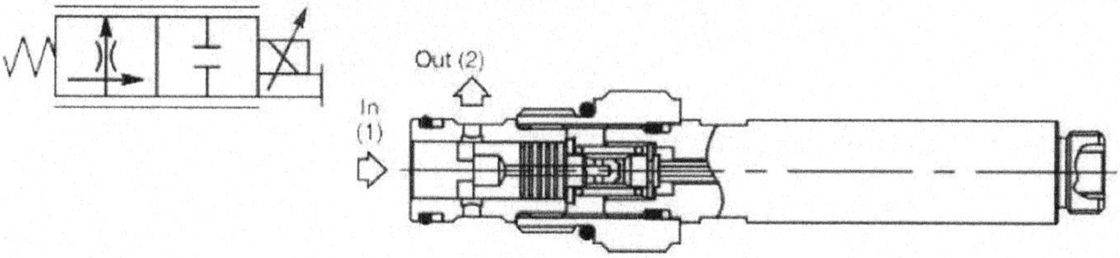

Fig. 2.14 Normally open proportional flow control valve [11]

The twin spool CMA valve from EATON is designed with anti-cavitation valves, pressure compensated metering in and metering out control and integrated with sensors for fault diagnosis. The valves claim a step response of 4 milliseconds, frequency response of 17.5 Hz (Fig. 2.15).

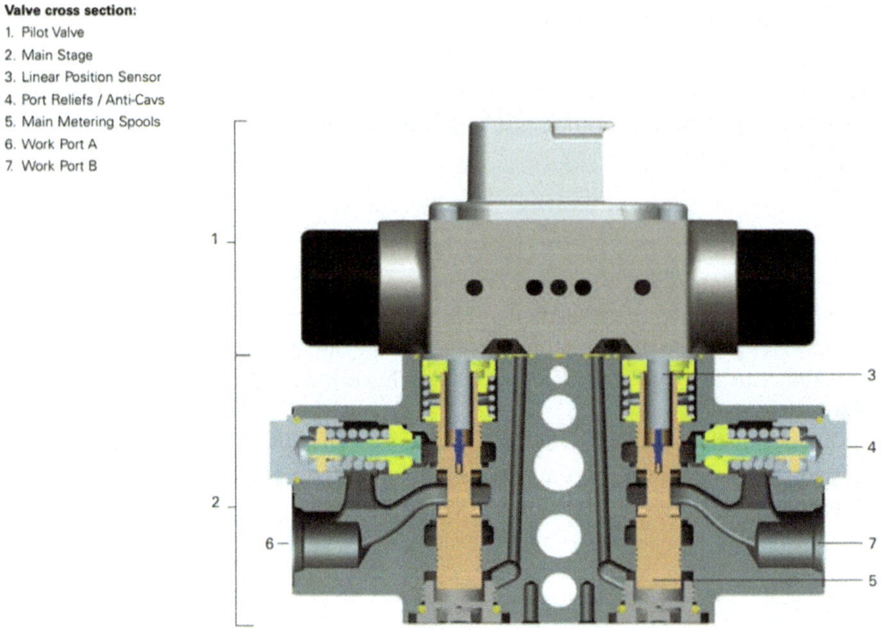

Fig. 2.15 EATONS twin spool CMA valve [12]

2.7 Linear Servo Actuators

Hydraulic cylinders that are fitted with low-friction Turcite seals serve as servo actuators. Vented rod bearings provide superior friction characteristics and permit operations at higher speeds and frequencies. Actuators with hydrostatic rod bearings exhibit good low-speed/frequency operations (Fig. 2.16 and Table 2.5).

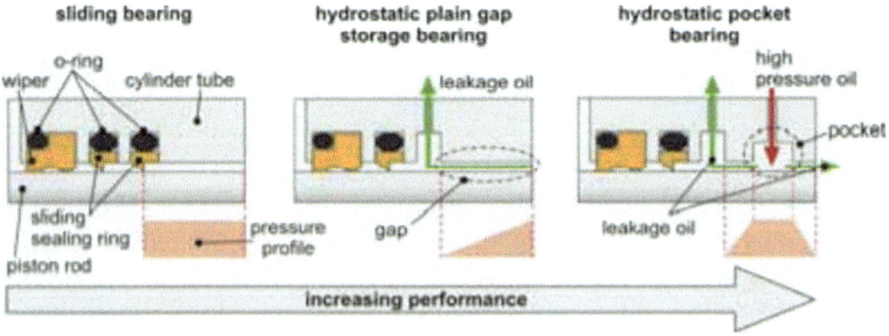

Fig. 2.16 Rod bearing types [10]

Table 2.5 Performance of three types of rod bearings [10]

	Industrial cylinders	High performance cylinders	Servocylinders
Bearing	Sliding bearing	Hydrostatic plain gap storage bearing	Hydrostatic pocket bearing
Transverse load capacity	Restricted	Restricted	High capacity
Friction	Dependent on pressure and transverse load	Independent on pressure, dependent on transverse load	Independent on pressure and transverse load
Max. operation frequency (Hz)	5	50	Limited by control valve
Max. velocity (m/s)	0.5	2	5
Position transducer	Magnetostrictive	LVDT or magnetostrictive	LVDT or magnetostrictive

Servo actuators closely couple servo valves with hydraulic cylinders and integrate the position transducer. The other functional valves necessary for the system are manifold between valve and cylinder. The supply pressure operated lock up slice aids in holding the load in the power off condition. Relief valves protect the cylinder from overpressures, and an emergency solenoid valve works in case of servo valve failure.

Integration is now being extended to include amplifiers and sensors to facilitate control/condition monitoring and troubleshooting (Fig. 2.17).

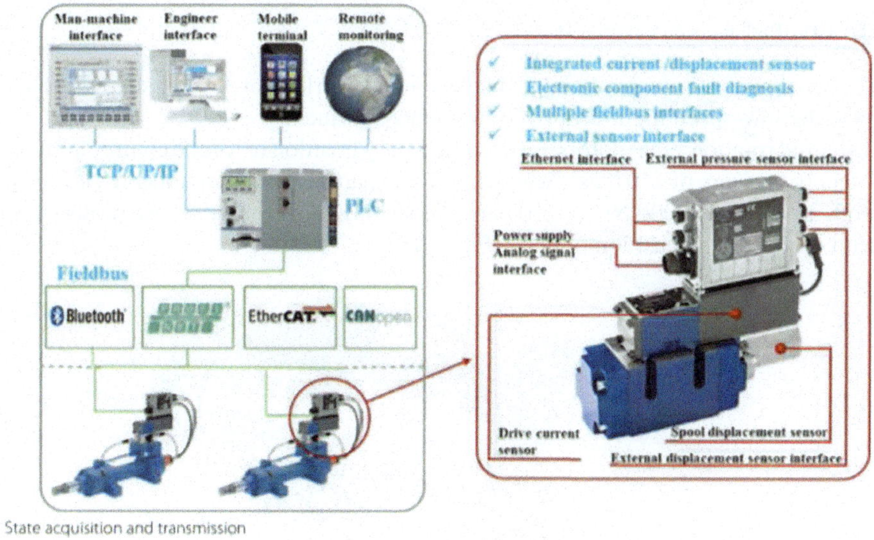

Fig. 2.17 Integrated actuator with sensors for monitoring [13]

2.8 Servo Drive Packages

For rotary position/speed control, EH servo drive packages close-couple hydraulic motors and servo valves with other functional valve manifolds. Axial piston and radial piston motors are combined with servo valves to form servo drive packages (Fig. 2.18).

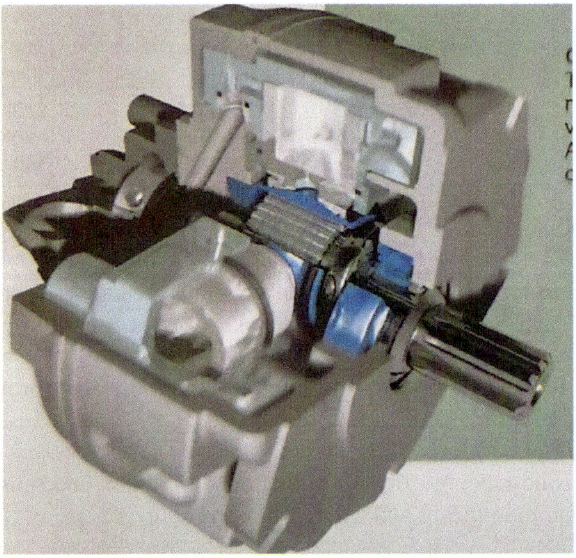

Fig. 2.18 Radial piston motor [14]

Radial piston motors with large displacements working at 350 bar supply pressure meet the low-speed high-torque requirements of winches.

Partial rotation vane motors are used for rotary position or torque control limited to a supply pressure of 140 bars. Geared hydraulic partial rotationmotors provide torque from 200 Nm to 32000 Nm at 210 bars cited by HKS, Germany. These motors come with encoders and cushion options (Fig. 2.19).

Fig. 2.19 Geared motors [15]

Rotary actuators derive rotation through racks, single/dual, and have a wide range of torque capability [16] (Fig. 2.20).

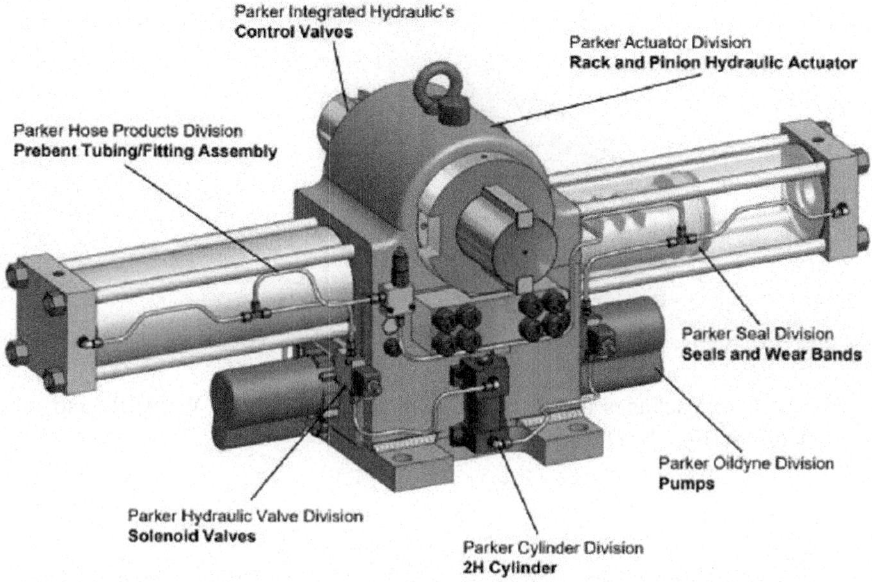

Fig. 2.20 Rotary actuators [16]

2.9 EHSA Drive Packages

EHSA drives extend integration to include pump, piston, or internal gear types but exclude any control valve. Some options from Voith are shown in Fig. 2.21 [17].

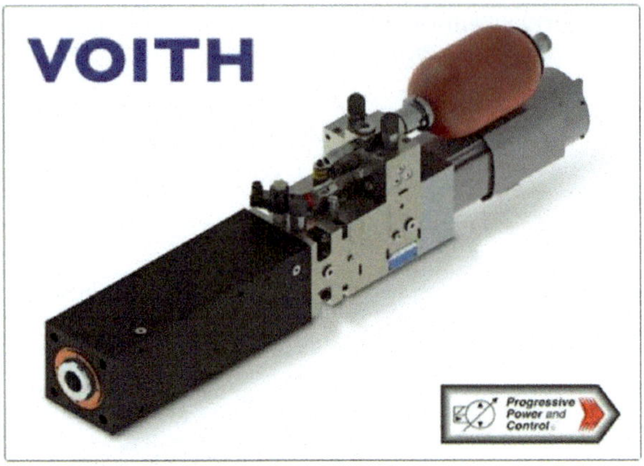

Fig. 2.21 EHSA drive packages [17]

EHSA drive packages are also available from Kawasaki, WEBER, and other manufacturers (Fig. 2.22).

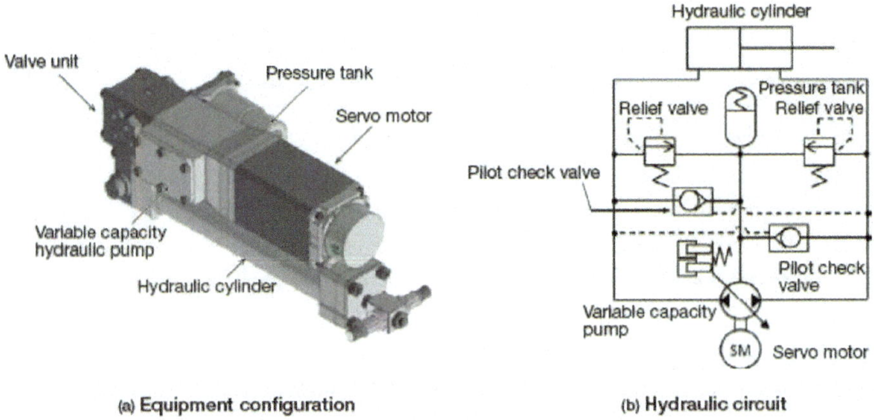

Fig. 2.22 EHSA Kawasaki [18]

2.10 High-Power Servo Motor-Driven Pumps

A VFD or AC servo motor driving a variable displacement pump to provide flow and pressure to match the load/flow demand is now introduced for higher-power ranges by Rexroth [19] (Fig. 2.23).

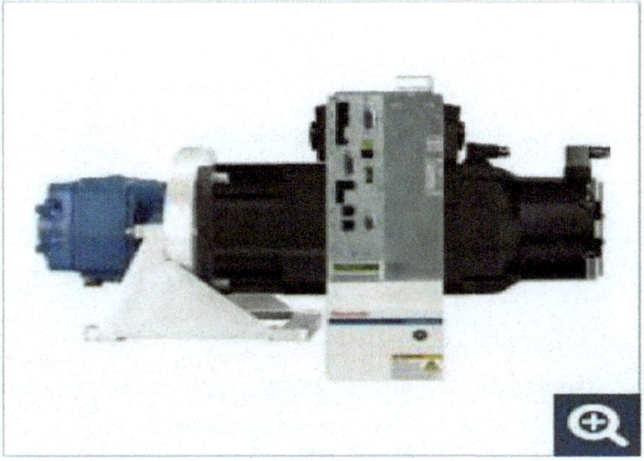

Fig. 2.23 Servo motor-driven pump [19]

2.11 Transducers

Control entails measurement, and transducers provide the measurement for EH servo control. They are picked for their accuracy, absolute, and repeatability. Transducers specify these in terms of linearity and hysteresis. Additionally, the response of transducers must be far higher than that of power elements that inhibit system performance.

2.11.1 Linear Position

A range of analog linear position transducers, potentiometers, LVDTs, DCDTs, and thermosonic transducers is used based on system needs.

LVDT is an inductive transducer that converts mechanical displacement to an AC electrical output signal. The working principle and output characteristics are shown in Figs. 2.24 and 2.25. They have infinite resolution and tend to be nonlinear around the null region [20].

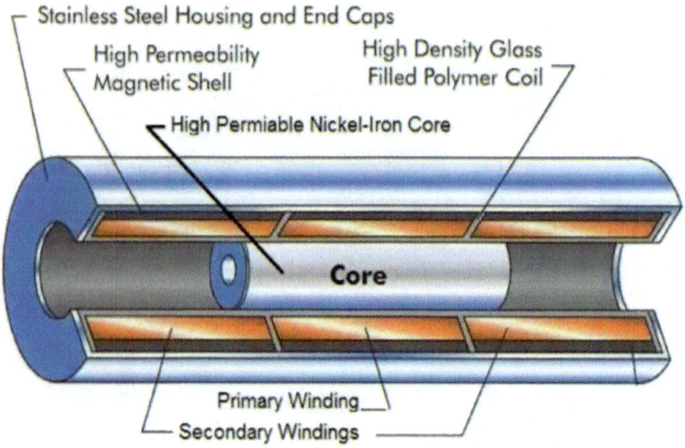

Fig. 2.24 Basic construction of the LVDT [20]

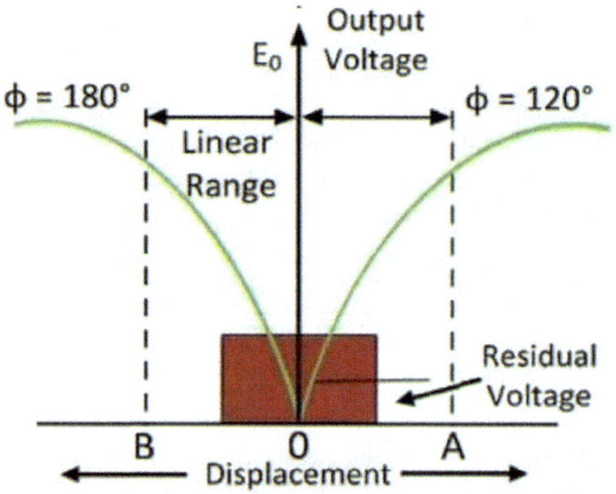

Fig. 2.25 Input/output of the LVDT [20]

They are available in ranges of ± 0.25 to ± 50 mm. The dynamic range is from 50 to 20 kHz [20] (Fig. 2.26).

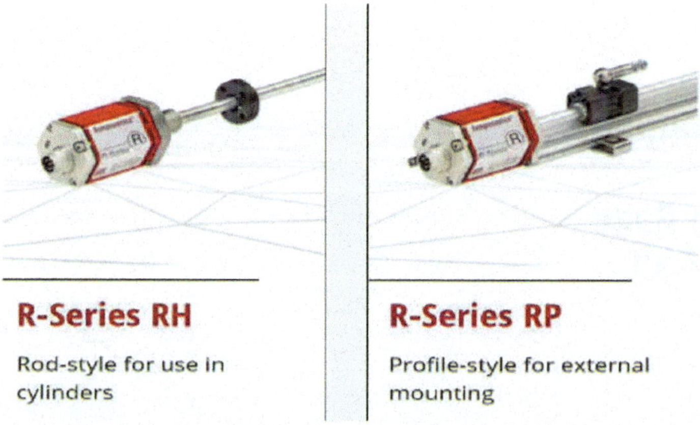

Fig. 2.26 Temposonic transducer [21]

Temposonic transducers working on the magnetostrictive principle, output analog or digital signals, and certain models can also provide velocity outputs. These are available over a wide range of strokes [21].

2.11.2 Load Cells

In addition to accuracy and response, load cells for EH servo application need to have overload capacity and must be fatigue rated. Load cell calibration is facilitated by shunt resistances [22] (Figs. 2.27 and 2.28).

Fig. 2.27 Load cells [22]

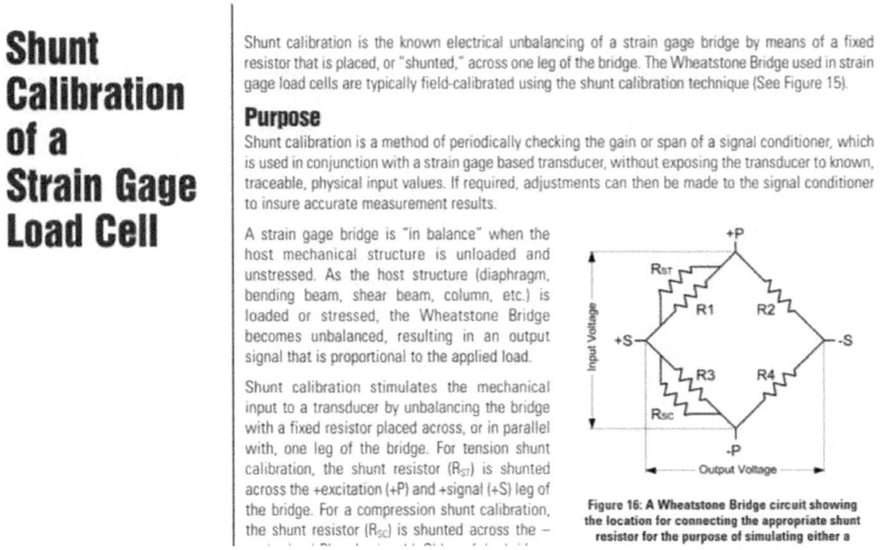

Shunt Calibration of a Strain Gage Load Cell

Shunt calibration is the known electrical unbalancing of a strain gage bridge by means of a fixed resistor that is placed, or "shunted," across one leg of the bridge. The Wheatstone Bridge used in strain gage load cells are typically field-calibrated using the shunt calibration technique (See Figure 15).

Purpose

Shunt calibration is a method of periodically checking the gain or span of a signal conditioner, which is used in conjunction with a strain gage based transducer, without exposing the transducer to known, traceable, physical input values. If required, adjustments can then be made to the signal conditioner to insure accurate measurement results.

A strain gage bridge is "in balance" when the host mechanical structure is unloaded and unstressed. As the host structure (diaphragm, bending beam, shear beam, column, etc.) is loaded or stressed, the Wheatstone Bridge becomes unbalanced, resulting in an output signal that is proportional to the applied load.

Shunt calibration stimulates the mechanical input to a transducer by unbalancing the bridge with a fixed resistor placed across, or in parallel with, one leg of the bridge. For tension shunt calibration, the shunt resistor (R_{ST}) is shunted across the +excitation (+P) and +signal (+S) leg of the bridge. For a compression shunt calibration, the shunt resistor (R_{SC}) is shunted across the –

Figure 16: A Wheatstone Bridge circuit showing the location for connecting the appropriate shunt resistor for the purpose of simulating either a

Fig. 2.28 Load cell calibration [22]

2.11.3 *Pressure Transducers*

Strain gauge-based pressure transducers are generally used to measure load pressure across the piston head for control purposes. Individual pressure transducers with matched amplifiers are manifolded between the servo valve and actuator. Pressure transducers with 0.25% FSD accuracies are common.

All strain gauge-based sensors need output amplification.

2.11.4 *Rotary Position*

Potentiometers, RVDTs, and encoders and synchros are used for angular position measurements. Synchros are relatively more accurate, rugged, and are used in analog control systems. Synchro-to-digital converters are required for their use in digital systems. Encoders with high resolution are available in incremental, absolute versions. Encoders are susceptible to temperature and mechanical shocks.

High resolution encoders are used for deriving state variables for control.

2.11.5 Rotary Speed

Speed measurements are performed with tachometers and are generally integrated with servo drive packages. When independently used, care should be taken in couplers to avoid backlash.

Speed range, linearity, and sensitivity are the factors critical for applications.

2.11.6 Other Sensors

Pendulum-based level sensors have been used for leveling systems.

Inertial measuring units (IMUs) report accelerations, angular rates, and other gravitational forces. It combines three accelerometers, three gyroscopes, and, depending on heading requirements, three magnetometers. MEMS IMUs combine high performance with ultralow power in a smaller envelope.

They find applications in systems demanding stabilized platforms (Fig. 2.29).

Fig. 2.29 MEMS IMU [23]

2.11.7 *Torque Transducers*

Torque transducers are available in different mounting styles, reaction types, and slip ring-based and now with telemetry options for a wide range of rated capacities. Most are strain gauge based needing amplification and have to be fatigue rated with overload capacity. Accuracies of 0.15% are quoted (Fig. 2.30).

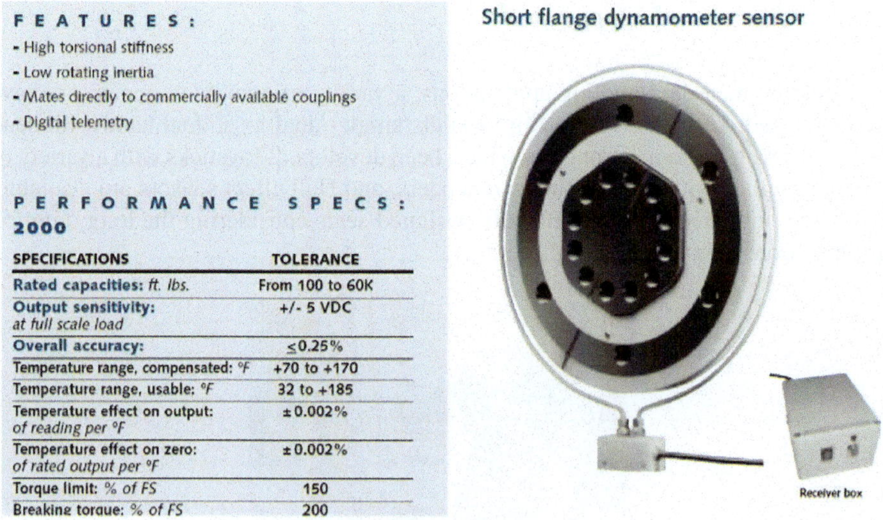

FEATURES:

- High torsional stiffness
- Low rotating inertia
- Mates directly to commercially available couplings
- Digital telemetry

PERFORMANCE SPECS: 2000

SPECIFICATIONS	TOLERANCE
Rated capacities: *ft. lbs.*	From 100 to 60K
Output sensitivity: *at full scale load*	+/- 5 VDC
Overall accuracy:	≤0.25%
Temperature range, compensated: *°F*	+70 to +170
Temperature range, usable: *°F*	32 to +185
Temperature effect on output: *of reading per °F*	±0.002%
Temperature effect on zero: *of rated output per °F*	±0.002%
Torque limit: *% of FS*	150
Breaking torque: *% of FS*	200

Fig. 2.30 Torque transducer with telemetry [22]

Telemetry solutions cater for data and power transfer or for data transfer only requirements [22].

Another development is the use of wireless transmission for data transfer of parameter data to a mobile monitor for fault tracing and control (Fig. 2.31).

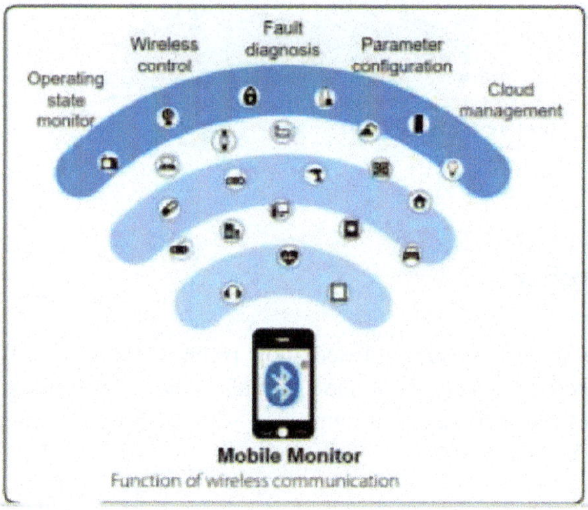

Fig. 2.31 Data transfer to mobile monitor [13]

2.12 Command Generators

For mobile, material handling applications, a variety of command generators have been devised. Manual control with joystick, single-/dual-axis, dual hand with dead man's control, and remote operation have been developed. Joysticks with a variety of grips, designed with potentiometers/encoders, and Hall effect sensors are available. These are mounted on ergonomically designed seats considering the long duration of operations (Fig. 2.32).

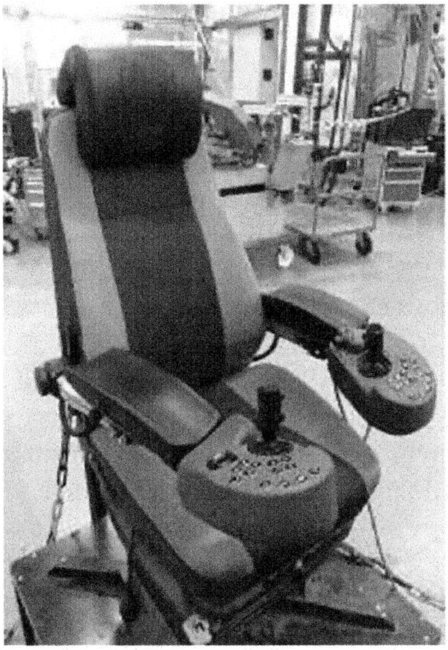

Fig. 2.32 Joysticks for forestry cranes [24]

2.13 Controllers

Controllers for analog systems closed loop provide for error PID manipulation and are designed to accept command signals. These incorporate signal conditioning modules for feedback and command signals. Single-/multiaxis controllers are designed (Fig. 2.33).

Fig. 2.33 Multiaxes servo controller [25]

Controllers for digital systems have been built with microprocessors and DSPs and are now invariably with computers. The exponential power of computing has been extended to meet the loop manipulations and the preloop and peripheral requirements of systems.

Hardware capabilities have been raised with FPGA development. Software meant for digital control design/analyses is at hand with MATLAB compatible software. Compact configurable systems are available for multiaxes systems.

References

1. Merritt HE (1967) Hydraulic control systems. Wiley
2. Moog Catalog. Jet Pipe Servo Valve. CDS33551. http://www.moog.com/industrial/. Updated 10 Sept 2022
3. Moog catalog. D765 series. http://www.moog.com/industrial/. Accessed 9 April 2021
4. Moog catalog. 72 series. http://www.moog.com/industrial/. Updated 10 Sept 2022
5. Plummer A (2016) Electro hydraulic servo valves-past, present and future. In: Proceedings of the 10th international conference
6. MTS 256 Servo Valve Catalog. 256_servovalve.pdf. http://www.mts.com/. Accessed 24 Dec 2021

7. Yuken Hydraulic Equipment. Servo valves. Catalog. http://www.yuken-usa.com/. Accessed 21 Dec 2021
8. Parker catalog. BUL HY14-2554-MI_DX1FP.pdf. http://www.parker.com/hydraulicvalve. Accessed 24 Dec 2021
9. Tambarrano P, Plummer AR (2019) A review of electro hydraulic servo valves. Research and development. Int J Fluid Power 20(1):53–98
10. Reichert M (2010) Development of high response piezo servo valves for improved performance of electro hydraulic servo drives. Shaker Verlag, GMbH, Germany
11. Parker Catalog. HY15–3502/us. http://www.fluidpowersolutions.ca. Updated 13 Sept 2022
12. EATON catalog.eaton-cma-90-mobile-valves-technical-catalog.pdf. http://www.eaton.com/hydraulics. Accessed 6 Jan 2022
13. Xu B, Shen J, Liu S et al (2020) Research and development of electro hydraulic control valves oriented to industry 4.0: a review. Chin J Mech Eng 33(29). https://doi.org/10.1186/s10033-020-00446-2
14. Catalog. SAI s.p.a. http://web.sai-hyd.com. Accessed 18 April 2021
15. Catalog. HKS. https://www.hks-partner.com. Accessed 14 Feb 2022
16. Catalog. Parker. HY-03-1800/us hydraulic rotary actuators. http://www.parker.com. Accessed 14 Feb 2022
17. Servo Drive CLDP Technical Data Sheet. VOITH, Germany. http://www.voith.com/hydraulic-systems. Accessed 14 Feb 2022
18. Kawasaki Catalog EHSA. http://global.kawasaki.com/. Updated 14 Sept 2022
19. Rexroth/Bosch. Sytronix. Variable speed drives. Catalog. http://boschrexroth.com/sytronix. Accessed 14 Feb 2022
20. Electrical-Technology.com <2019/05> LVDT. Accessed 10 Jan 2022
21. Catalog. MTS Temposonic Transducers. MTS. www.temposonics.com. Updated 14 Sept 2022
22. LeBow Products.Honeywell.LeBow catalog 2006. http://lebow.com. Accessed 10 April 2021
23. Catalog. HG4930 MEMS IMU.Honeywell. http://aerospace.honeywell.com. Accessed 13 Jan 2022
24. Fodor S (2017) Towards semi-automation of Forestry Cranes. UMEA University, Sweden
25. Moog Model 127-101. Motherboard controller. http://controller.moog.com/industrial. Accessed 14 Feb 2022

Chapter 3
Applications of EH Servo Systems

Abstract The gamut of engineering applications in the engineering industry using EHSS are recalled under Manufacturing, Mobile, Motion Simulators, Material & Structure Testing, Robots, Defence, Aircraft & Aerospace and others and described.

The power and quadratic response of EH servo systems have made them a popular choice of drive for a wide range of engineering applications. The recent developments in EHSA technology vying for efficiency with electromechanical systems have extended this range. Some of the representative applications are related industrywise here.

3.1 Manufacturing

3.1.1 Friction Welding

Automotive axles and cylinder rod ends are rotary friction welded as the joint is stronger than the parent material strength. The process initiates by rubbing with a light friction force to remove impurities. The spindle speed is increased against increased friction force to the desired rubbing velocity, and as the material melts and the desired burn-off is obtained, the spindle is stopped. A swaging force approximately double the friction force is then suddenly applied to forge a joint. Machines are rated for swaging force depending on the area of rubbing and material. The system needs to work in position and force modes and switch at the right time [1] (Fig. 3.1).

D. R. Raghavendra, *Electrohydraulic Servo Systems*, Springer Tracts in Mechanical Engineering, https://doi.org/10.1007/978-981-19-8065-7_3

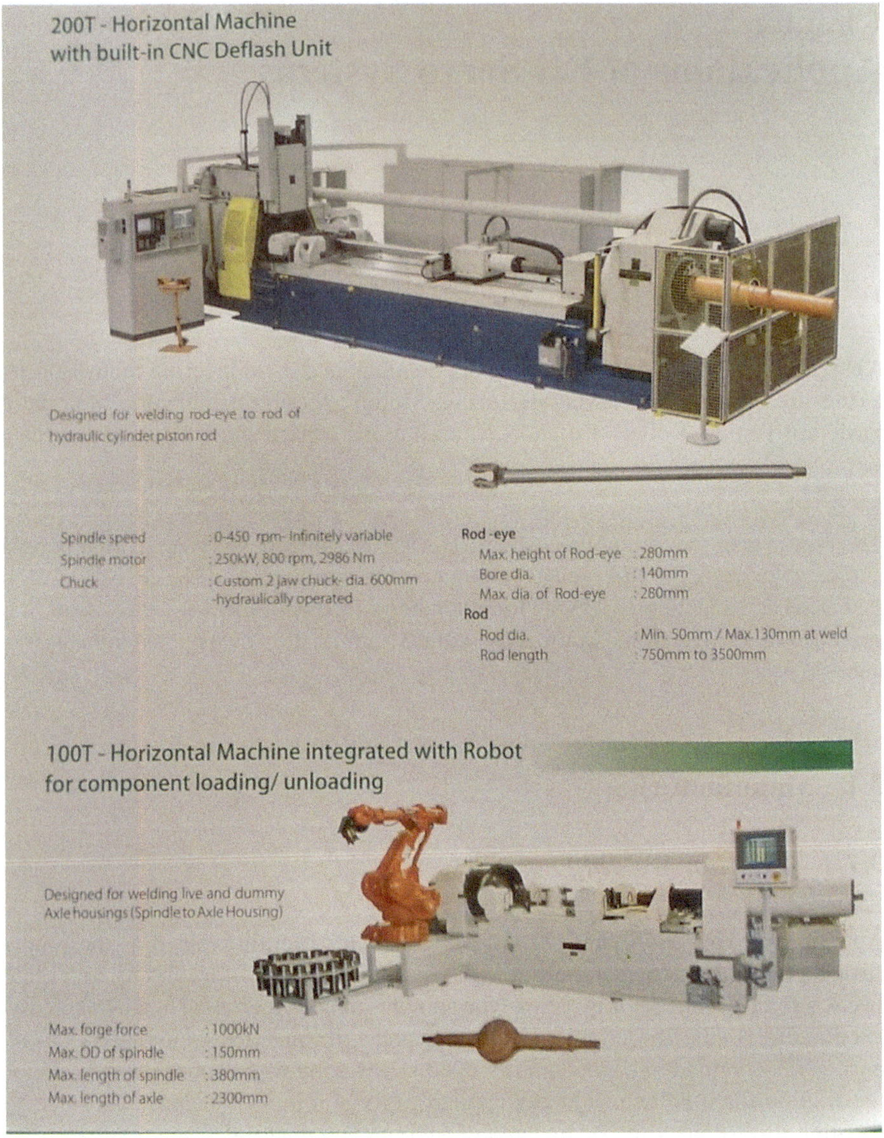

Fig. 3.1 Rotary friction welding machines [1]

Robot integration to handle work pieces increases machine throughput (Fig. 3.2).

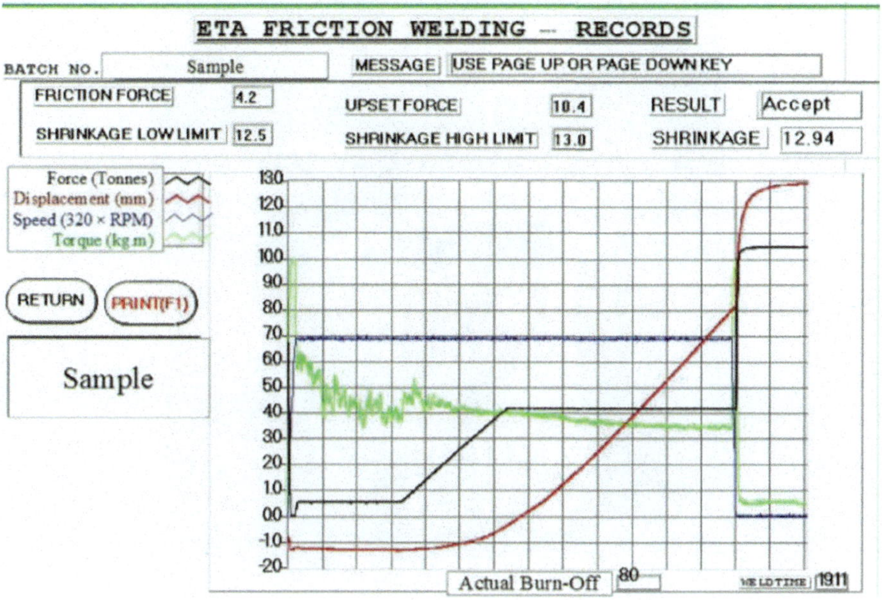

Fig. 3.2 Friction welding process [1]

3.1.2 *Injection Molding*

Injection molding is one of the oldest applications that call for mode switching from position to force. While initial filling of mold requires good velocity control, switching to force control at the opportune moment yields good strength to the molded component. The requirements of pressure, velocity (flow), temperature over time, and distance have been met by many designs of valve-controlled and pump-controlled EHSA drives. These machines of varying capacities, in tonnage, are made by several manufacturers (Fig. 3.3).

Fig. 3.3 Injection molding machine [2]

3.1.3 *Blow Molding*

A popular application of EH servo position control is in blow molding machines. The die gap determines the shape of the blow molded component. The operator sets the desired shape of the component on a Parison controller (Fig. 3.4).

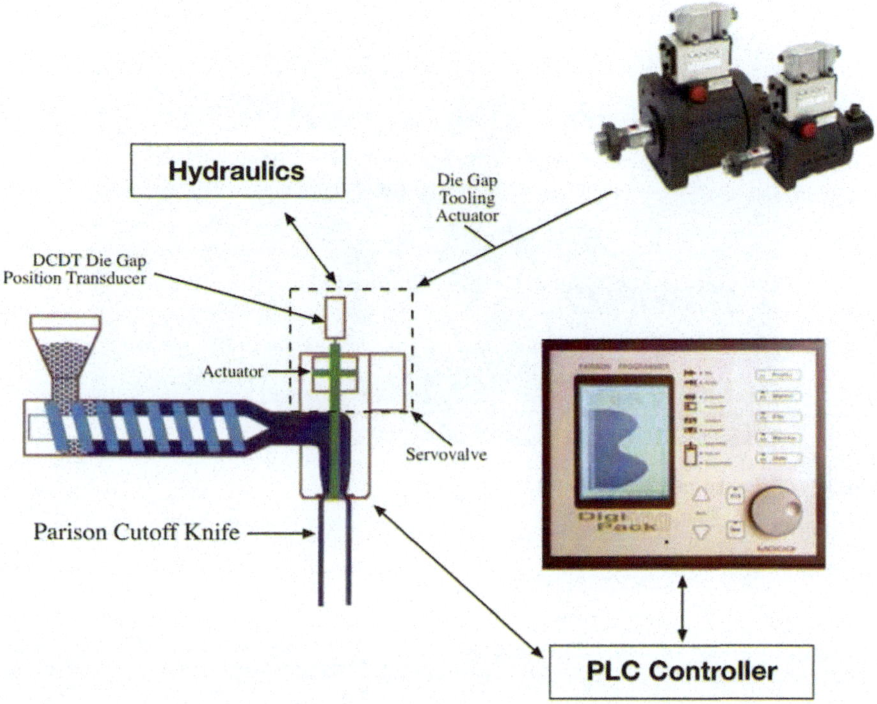

Fig. 3.4 Blow molding machine [3]

3.1.4 Globoidal Cam Fixture

Globoidal cams are part of Indexing units that provide fast, jerk-free, and smooth indexing. Machining of indexing cam [4], a candidate for electronic gearing of a two-axis drive, was realized in a setup with a hydraulic index table drive (Figs. 3.5 and 3.6).

Fig. 3.5 Globoidal cam assembly [4]

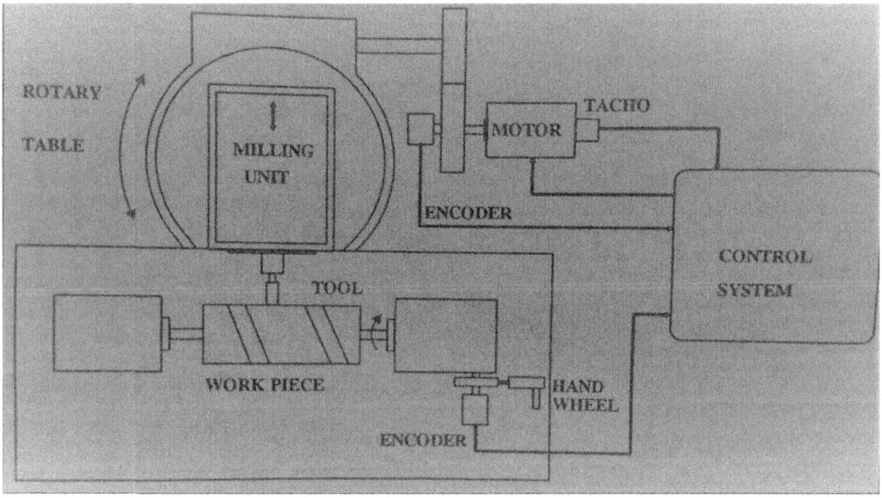

Fig. 3.6 Globoidal cam machining setup [4]

Based on the hand wheel position, the indexing speed varied smoothly with the required dwell (Fig. 3.7).

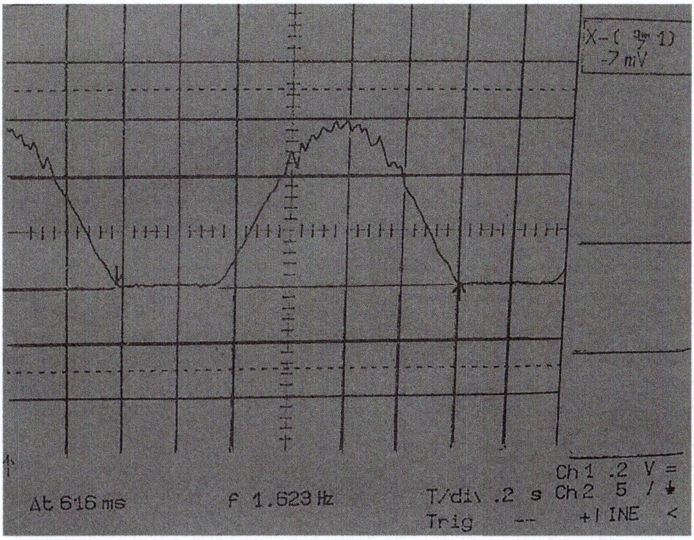

Fig. 3.7 Velocity profile of cam [4]

3.1.5 EHSA Hydraulic Press

The varying force/velocity requirements of a heavy press of 400 ton/0.03 m/s and 40 ton/0.27 m/s speed were met by an EHSA drive with a single pump. The design used a special 3-chamber actuator [5] (Figs. 3.8 and 3.9)

a) high moving mass.
b) discontinuous speed during adaption.
c) continuous process force.

+ heavy press, forming
+ cutting (high weight tool)
+ punching (low speed)

Parameter	Value
F_{MAX}	4 MN (400 to)
V_{FMAX}	0,03 m/s

Fig. 3.8 EHSA hydraulic press [5]

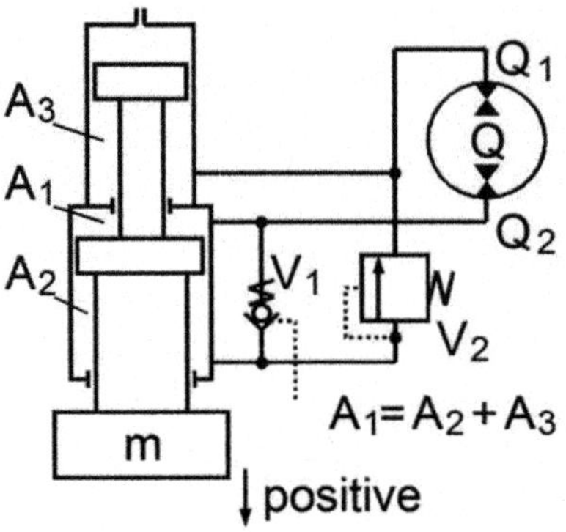

Mode	V1	V2	Q1	Q2
No load positive			A3	A1- A2
No load negative			A3	A1- A2
Full load positive			A2+ A3	A1

Fig. 3.9 Hydraulic circuit of press [5]

3.2 Mobile Earthmoving

Mobile Equipment: Heavy earth moving equipment is either tracked or wheeled with outrigger jacks. The functional requirements are met by jointed kinematics by a number of rugged actuators in the vertical plane and slewing at slow speeds and steering by hydrostatic drives. The need for matching the power of the IC engine source with that of hydraulics is a major design objective. Likewise, the efficiency of hydraulic operations is of prime importance for these applications.

3.2.1 Forestry Cranes

Automating forestry operations from logging, harvesting, and forwarding is achieved by forestry crane designs. The cranes are operated by two joysticks on either side of the driver's seat. The joysticks are assigned to operate slewing at low speeds in the horizontal plane and the other for actuations in the vertical plane [6] (Fig. 3.10).

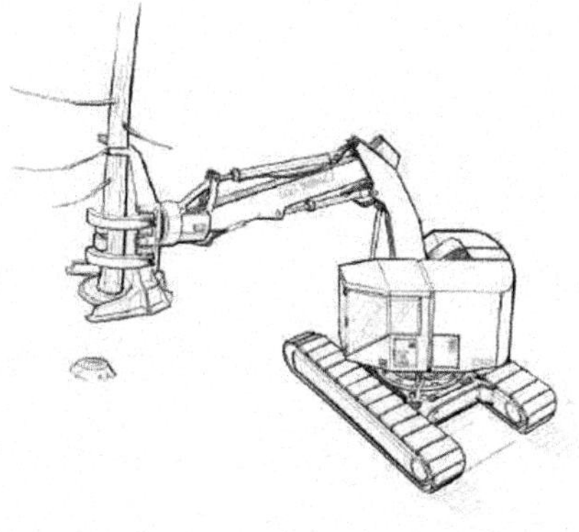

Feller-buncher.

Grapple skidder.

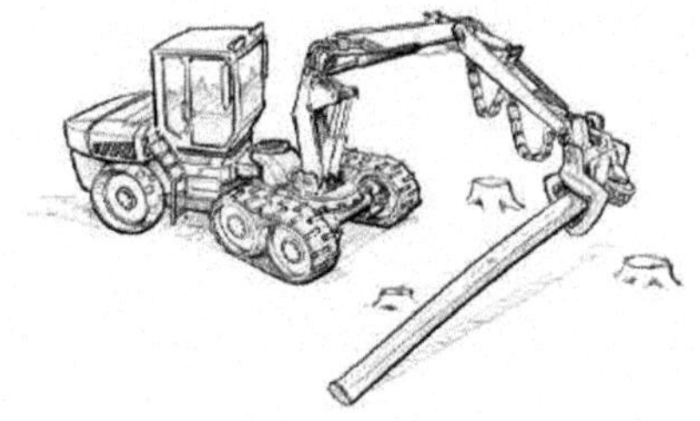

Single-grip harvester.

Forwarder.

Fig. 3.10 Hydraulic forestry cranes [6]

3.2.2 Hydrostatic Drives

Stepless variation of speed of hydrostatic transmission and the efficiency of mechanical power split transmissions are used to maximize use of available IC power. Two variants of mechanical power split transmissions are shown here (Figs. 3.11 and 3.12).

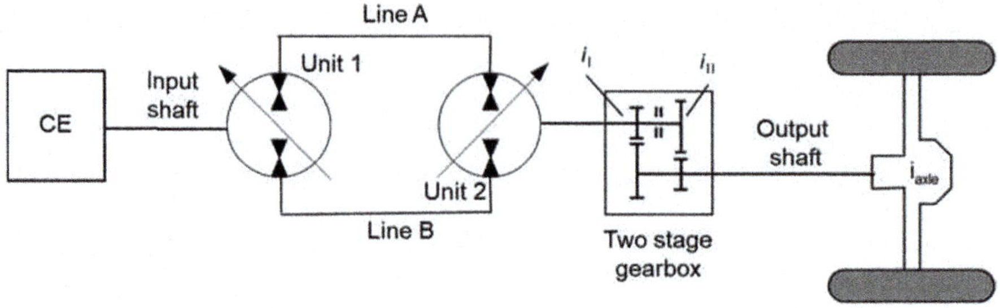

Fig. 3.11 Hydrostatic transmission in traction drives [7]

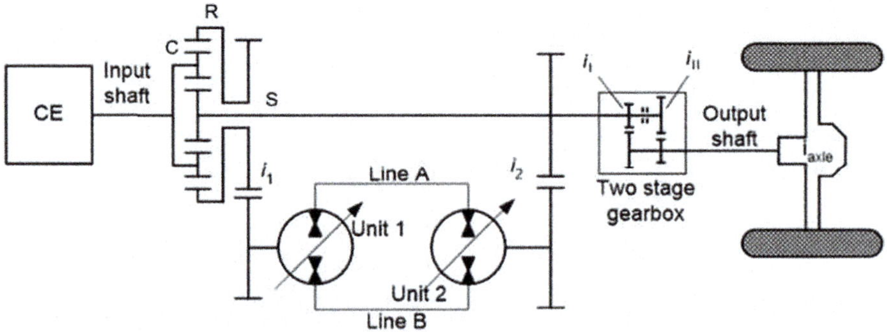

Fig. 3.12 Power transmission design variants [7]

The dual functions of drive and steering are obtained by differential steering of tracks in this design of a mobile hydrostatic drive [8] (Fig. 3.13).

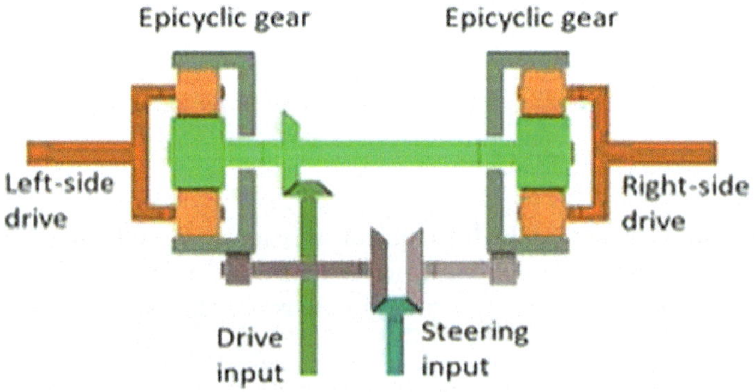

Fig. 3.13 Differential steering [8]

Proportional valve-operated systems with commands generated from stacked joysticks are used.

3.2.3 Power Steering

Power steering of heavy trucks is effected by the use of a rotary servo valve feeding actuator driving steering linkages. Power steering requires a sense of load to avoid overloading. This is done through a rack/pinion connected to a valve spool [9] (Fig. 3.14).

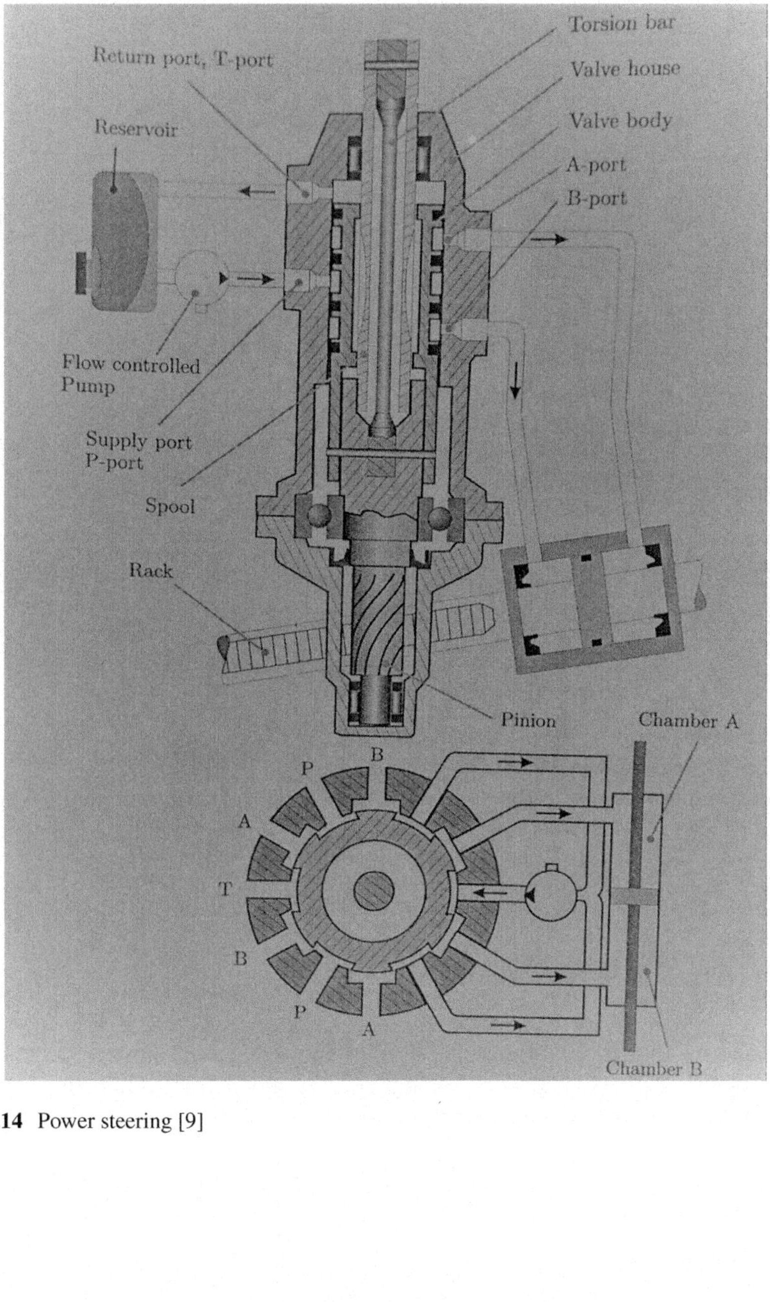

Fig. 3.14 Power steering [9]

3.3 Motion Simulators

Motion simulators form a main subsystem of flight simulators, Automobile Drive-Assist Systems (ADAS). They are configured with 4 or 6 legs in parallel linkage, whose lengths are varied for motion simulation.

3.3.1 Sea State Simulator

A sea state simulator is shown in Fig. 3.15 with a moving mass of 30 tons and speed reaching 1.5 m/s at an acceleration of 1.2 g.

Fig. 3.15 30-ton sea state motion simulator [10]

The simulator deploys 6 tandem servo actuators, each approximately 6 m long, fed with servo proportional valves with temposonic transducers for feedback.

The total power of the system is 1000 kW [10].

3.3.2 EHSA Flight Simulator

EHSA-driven motion flight simulators claiming substantial power reduction are shown in Fig. 3.16 [11]. Energy savings are realized by a combination of counterbalancing and avoidance of the servo valve in EHSA (Figs. 3.17 and 3.18).

Fig. 3.16 EHSA-based Boeing aircraft simulator [11]

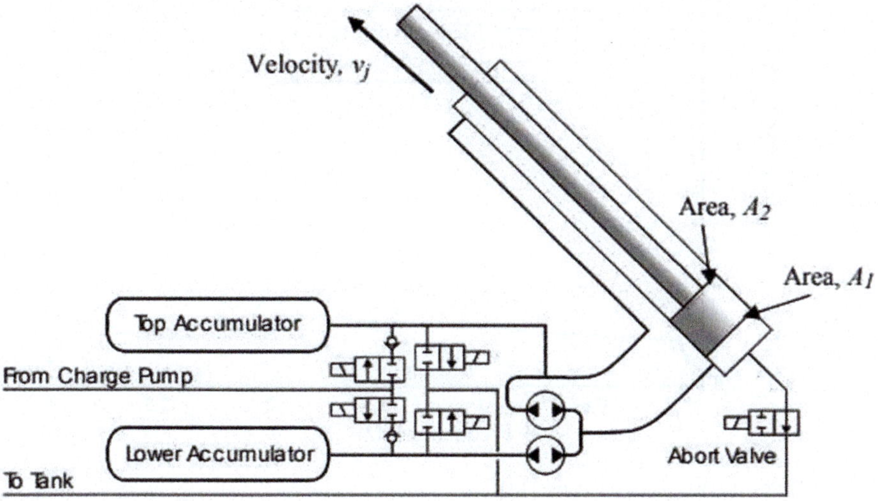

Fig. 3.17 EHSA circuit [11]

Platform/jack parameters		I. ornments
Piston head area, A *t*	86.6 cm^2	
Piston rod-side area, A,	42.4 eat:	
Jack over vertical motion ratio, *R*	0.787	
Conventional valve-controlled jack		
Supply pressure, P_c	*117* bar	
Water pump/leakage power, *'il*	10 kW	
Relief valve power loss, 'P$_r$	.-c. 8.5 kW	Maximum when static, rapidly becoming small with rising velocity
Drive/pump efficiency,/7,	0 85	
New electrobydrostatie system		
Quiescent power loss, Srf,"	4 kW	
Head side pressure loss coefficient, K$_i$	619x10'° mNs¯Nt·	Gives 22bar loss at 0A8mis jack velocity
Rod side pressure loss coefficienL K,	2.66x1& I m'/(s·N1	Gives 22bar loss at 0.48mis jack velocity
Effective platform Mass, *M*	27 400 *kg*	Includes II 400 kg drive/pump inertia referred to platform
Drivopump efficiency, r,	0,85	

System parameters€€rs

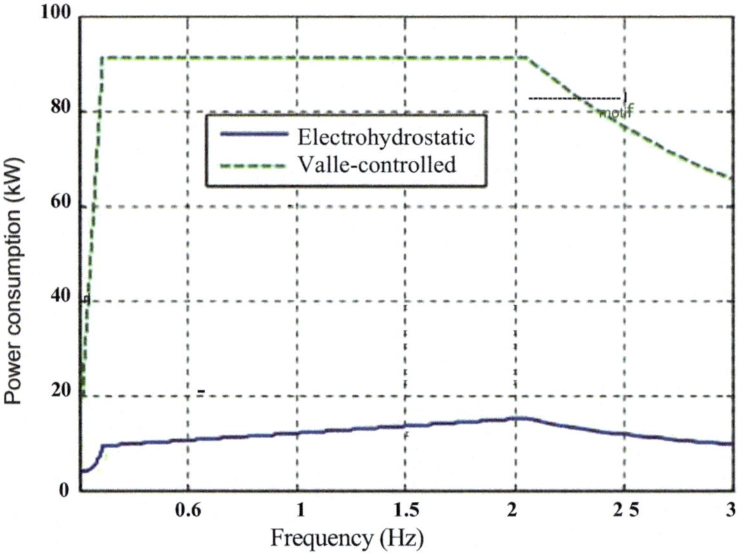

Fig. 3.18 Power savings of the drive [11]

Boeing 787 Dreamline simulators are powered by the EHSA system.

3.3.3 Automobile 4-Posters

Automobile 4-posters that simulate rough riding also use valve-controlled actuators [12] (Fig. 3.19).

Fig. 3.19 Road ride simulator [12]

3.3.4 Twist Rig

Railway standards demand multibody simulation of units consisting of 2 bogies, coaches, and primary and secondary suspensions going around specified curves. Unit under test undergoes a twist as it enters a curve on a cant, sways when on curve, and twists in reverse direction as it leaves the cant on to tangent section—a statutory requirement as part of derailment studies. Twist Rigs do so aided by 8 nos of servo actuators, whose coordinated motions enable derailment coefficients to be measured under loaded and unloaded conditions (Fig. 3.20).

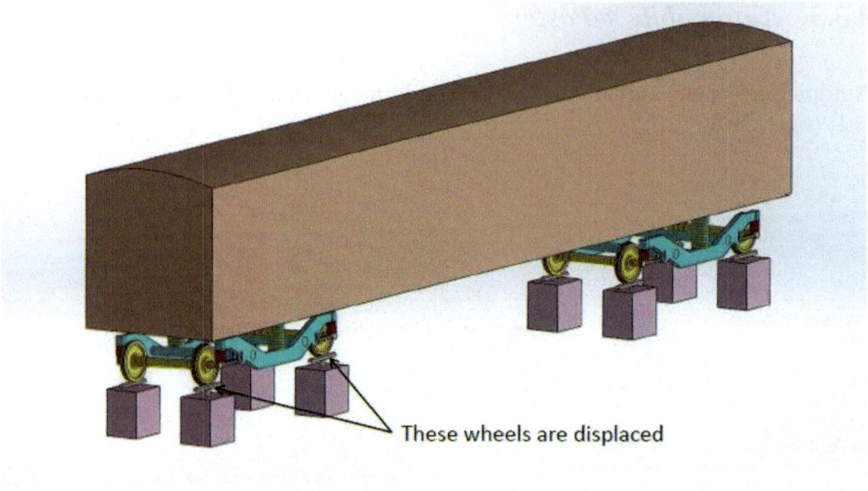

Fig. 3.20 Twist rig for rail derailment study [1]

3.4 Material and Structure Test Rigs

3.4.1 Landing Gear Actuators Test Rig

A universal end-of-line Test Rig for a family of aircraft landing gear actuators is shown in Fig. 3.21 [1]. Rig uses 2 reservoirs—one uses hyjet oil used on commercial aircraft, and the other uses mineral oil for loading test units. The structural integrity test for proof pressure to 950 bars uses a high-pressure check ball pump driven by a servomotor in a closed loop with a pressure transducer. The rig uses an asynchronous servo motor driving a variable delivery pump that enables precise measurement of actuator friction (<3 bar) and enables other tests demanding flows at higher pressures.

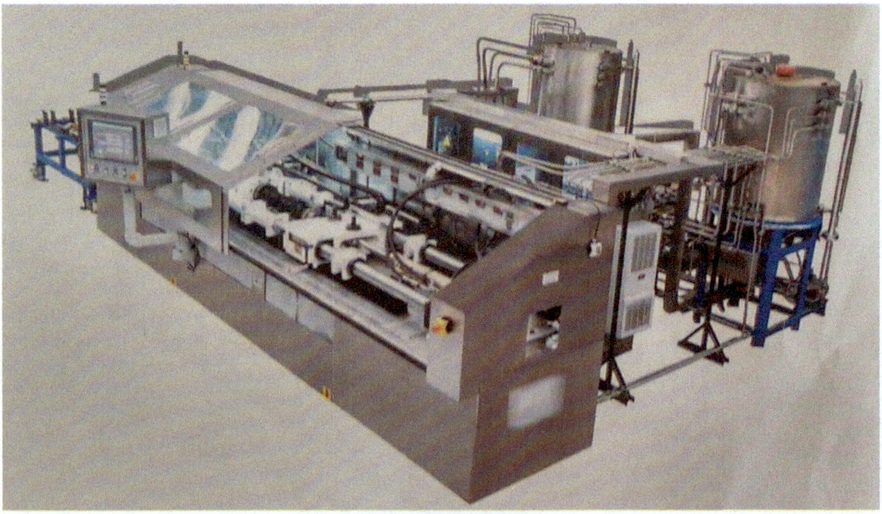

Fig. 3.21 Universal landing gear actuator test rig [1]

3.4.2 5-Axis Ball Joint Test Rig

Automobile ball joints need to be tested for endurance under multiaxes loading and articulation.

Range and adverse environmental conditions of dust and water. A 5-axis ball joint test rig from ETA technology is shown in Fig. 3.22. The design provides decoupled kinematics [1].

Fig. 3.22 5-Axis ball joint test rig for endurance and environmental tests [1]

3.4.3 3-Axis Steering Gear Test Rig

A 3-axis steering gear test rig for performance and endurance is shown in Fig. 3.23.

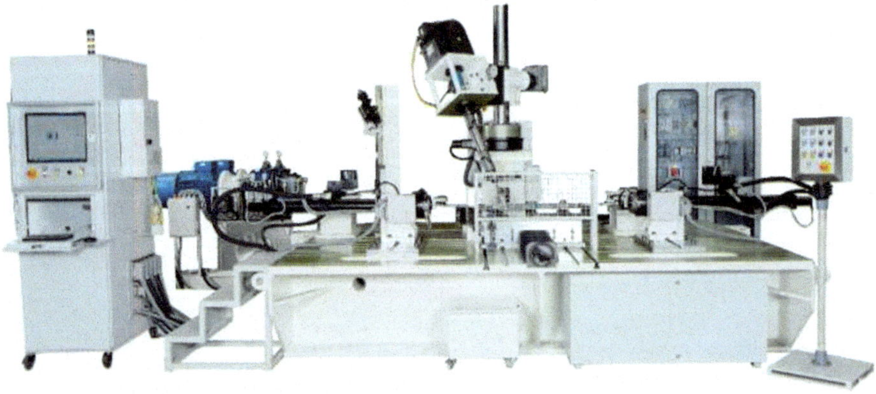

Fig. 3.23 3-Axis steering gear test rig [1]

Automobile manufacturers have rigorous test specifications for steering systems. System road loads are applied by EH servo actuators either directly or bell crank levers and steering action by an asynchronous servo motor [1].

All steering system designs can be tested with an environment chamber. The addition of two vertical cylinders enables jounce loads to be applied.

3.4.4 Aircraft Landing Shock Absorber Test Rig

The oleo-pneumatic shock absorbers are tested for performance and endurance and need to be periodically serviced by charging with air-free oil and pure nitrogen under pressure. A Universal Test Rig for both the main and nose wheel gears in most commercial aircraft is shown in Fig. 3.24 [13]. The rig facilitates mounting of gears by fixtures.

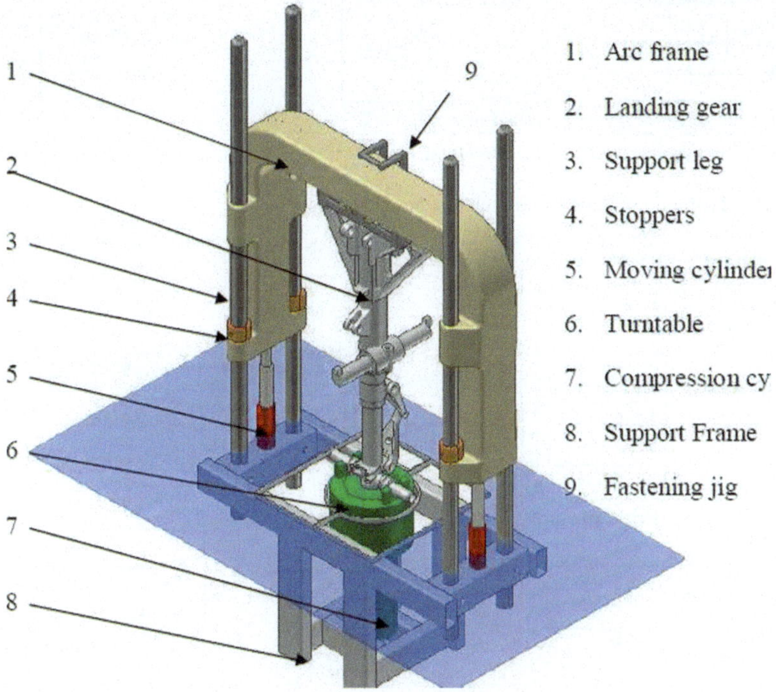

Fig. 3.24 Universal rig for aircraft landing gear shock struts [13]

3.4.5 Fatigue Test on Insulators

Insulators used in power transmission lines are required by law to be tested for fatigue under tensile and torsional loading. The high cycle fatigue test represents a ±10 ton tensile load at a frequency of 20 Hz, as shown in Figs. 3.25, 3.26, 3.27 [14] (Table 3.1).

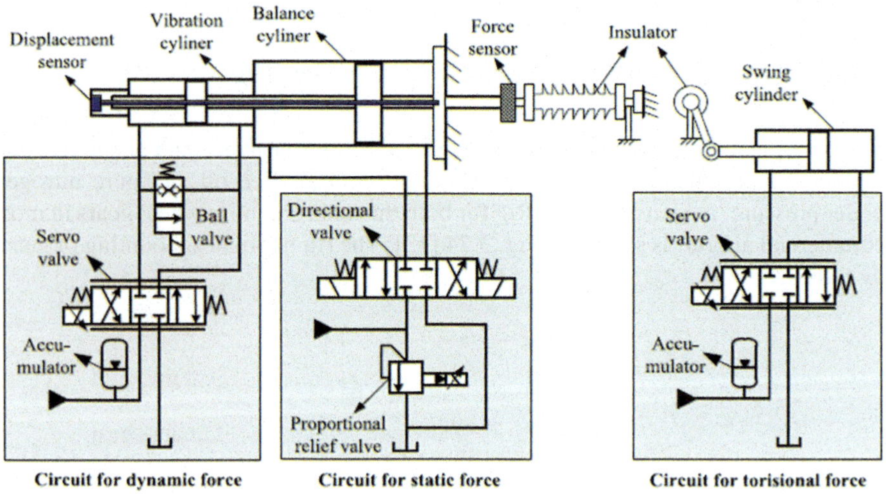

Fig. 3.25 Hydraulic scheme for fatigue test on insulators [14]

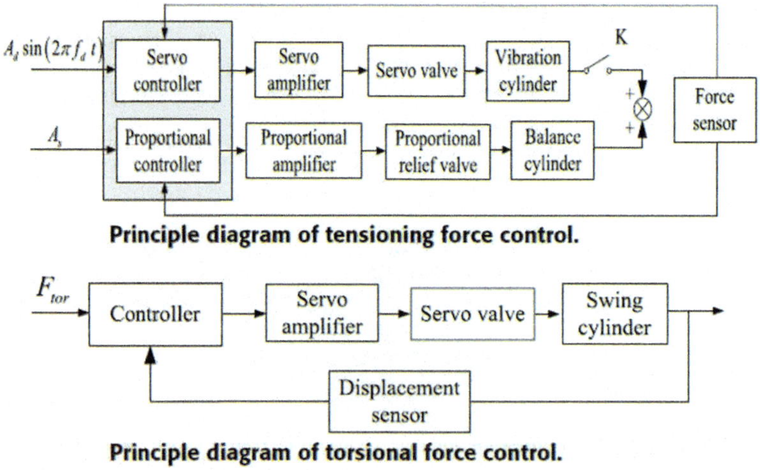

Fig. 3.26 Hydraulic scheme for fatigue test on insulators [14]

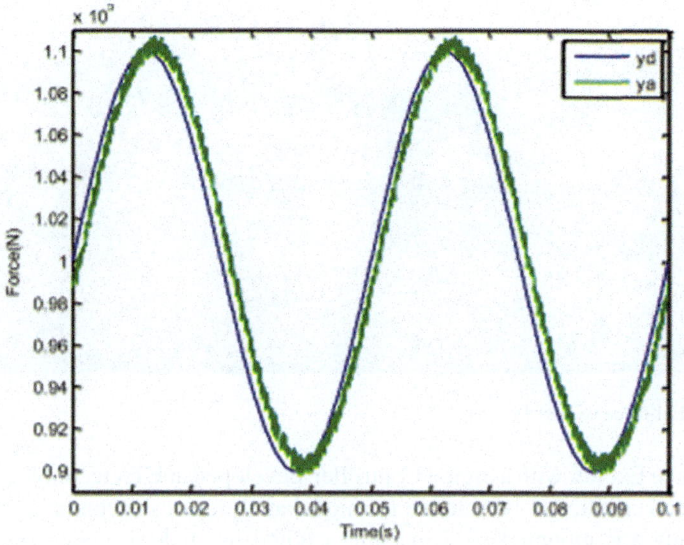

Fig. 3.27 Fatigue load tracking at 20 Hz [14]

Table. 3.1 Force tracking at frequencies [14]

f/Hz	$\mathbf{e}_{\mathrm{amp}}/N$	$e_{\mathrm{pha}}/0$
20	40	−10.2
40	150	−20.5
60	360	−30.9
80	680	−41.5
100	1120	−52.1

3.4.6 EH Force Exciters

Chatter in machine tool structures inhibits the power utilization and productivity of the machine and affects the surface finish of parts. Chatter is caused by a weak link in the structure, and modal analysis is carried out to identify this weak link with an ensemble of a force executor, triaxial accelerometer, and modal software. Force Exciter has to be compact and superpose dynamic forces over a static force over a large frequency range [15] (Fig. 3.28).

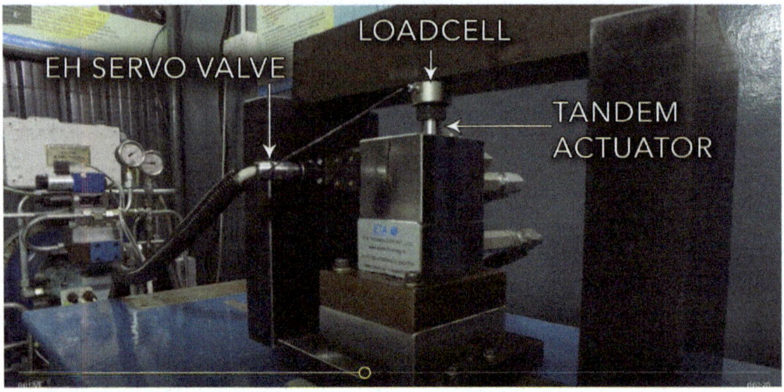

Fig. 3.28 EH force exciter [1]

EH Force Exciter with a digital Controller, developed at ETA technology, is tested, verified, and certified to be suitable for modal analyses of structures.

The unit was mounted on a machine excited in random mode. The structural response at multiple points was measured by a triaxial accelerometer. The modal parameters of natural frequencies and damping factors were extracted by Modal analysis software (Figs. 3.29 and 3.30).

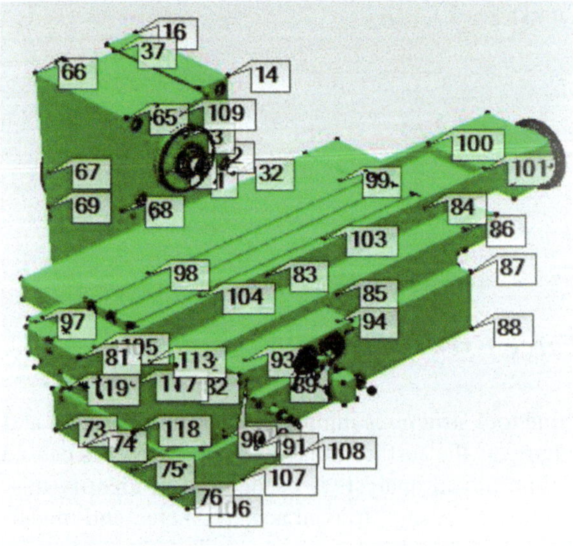

Fig. 3.29 Response measurements at multiple points [1]

Frequency	Damping in Hz	Damping (%)
11.6	0.285	2.46
17.5	0.793	4.53
39.9	1.48	3.71
44.9	1.24	2.76
52.4	2.71	5.16
101	4.41	4.35
112	5.23	4.65
181	8.23	4.54
216	2.71	1.25
284	5.91	2.08
316	5.32	1.68
338	12.8	3.8
409	9.58	2.34

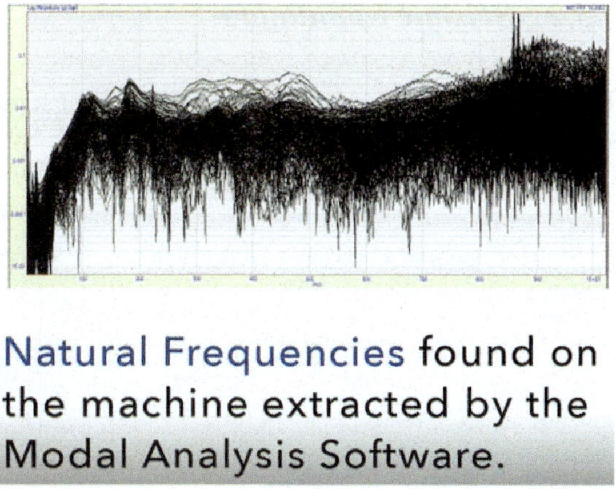

Fig. 3.30 Results of structural analysis [1]

3.5 Vibration Exciters

3.5.1 Vibration Table

EH vibration exciters are compact and provide large forces and velocities at frequencies below 1 kHz. The frequency/velocity performance of servo actuators is drawn on a log/log scale (Fig. 3.31).

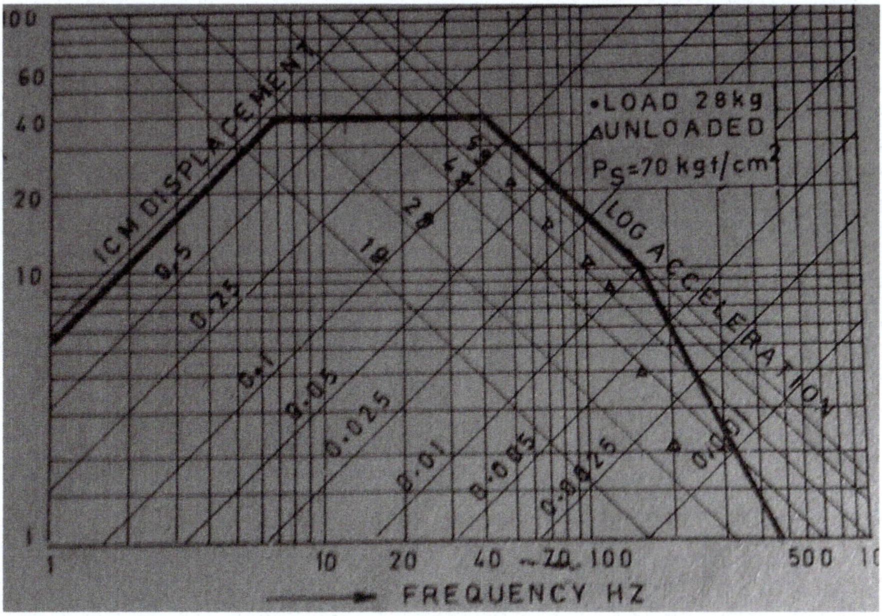

Fig. 3.31 Frequency versus velocity of vibration tables

3.5.2 Seismic Simulators

The effects of earthquakes on scaled models of superstructures are studied on seismic simulators. Seismic tests are performed either by acceleration tests or by effective force tests [16] (Fig. 3.32).

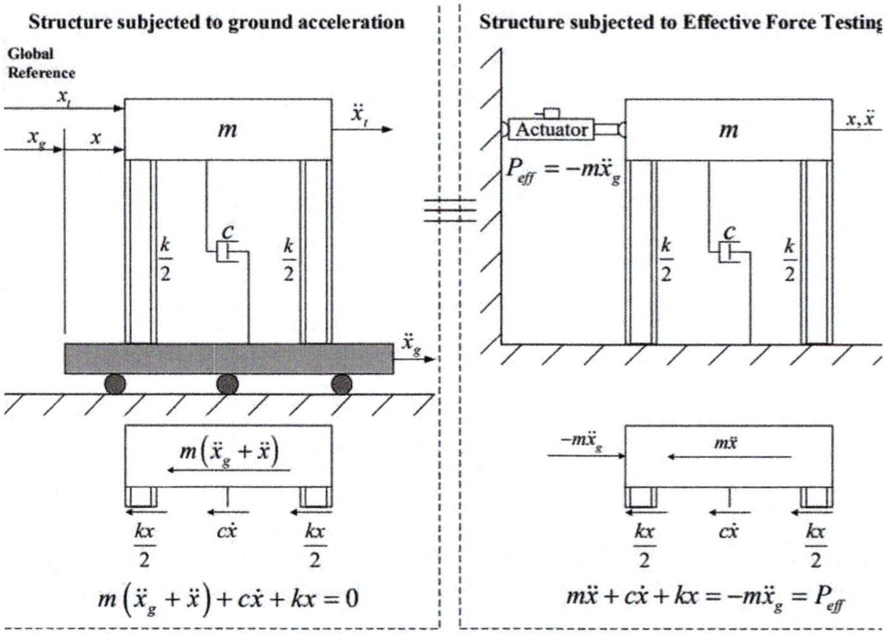

Fig. 3.32 Seismic test methods [16]

Acceleration tests are performed with slip tables with hydrostatic bearings. The controlled variable is acceleration. Effective force tests consider ground reactions and account for natural velocity feedback. Effective force tests demand large flows to sustain force around the natural frequency of the UUT. Uni- and multiaxial simulators are set up as national facilities (Fig. 3.33).

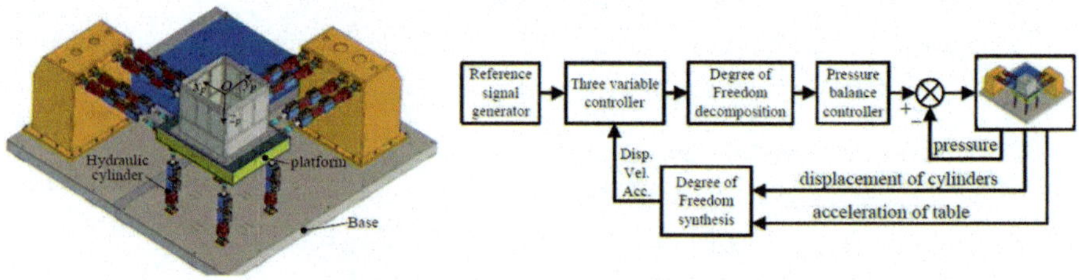

Fig. 3.33 6DOF redundant simulator [17]

3.6 Material Handling

3.6.1 300-Ton Critical Mass

The requirement was to pick/place a long slender critical object surrounded by a heavy shield in the x and y directions with a total moving mass of 300 tons to a desired accuracy of <1 mm in the x- and y-axes. 2 EH servo drives for the trolley in the y direction and 4 EH servo drives for carriage movement on rails were designed. Control for coarse positioning by NC commands. Fine positioning was obtained by a DCDT transducer closing loop [18].

3.6.2 Cranes, Loaders, Hoists

Most material handlers use proportional hydraulic servo systems, which are recommended by stacked joysticks (Fig. 3.34).

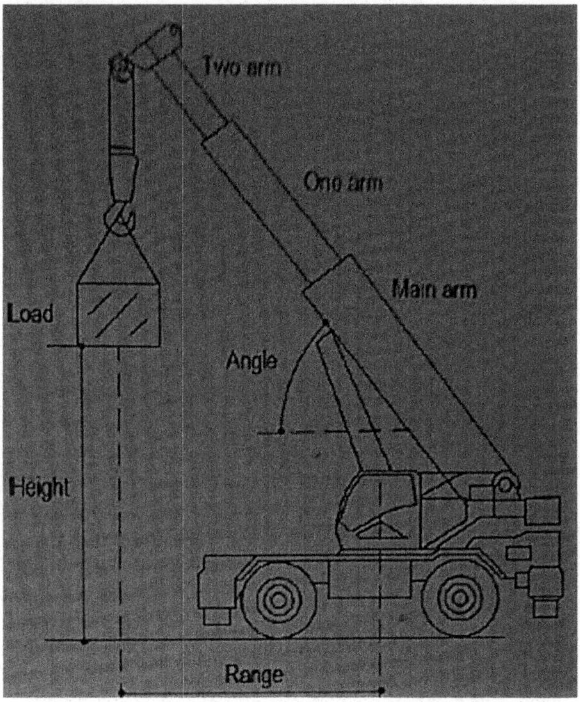

Fig. 3.34 Hydraulic crane [19]

To regenerate energy while unloading, these high-powered machines incorporate hydrostatic forces to drive a permanent magnet generator to charge the battery through IGBTs [19] (Fig. 3.35).

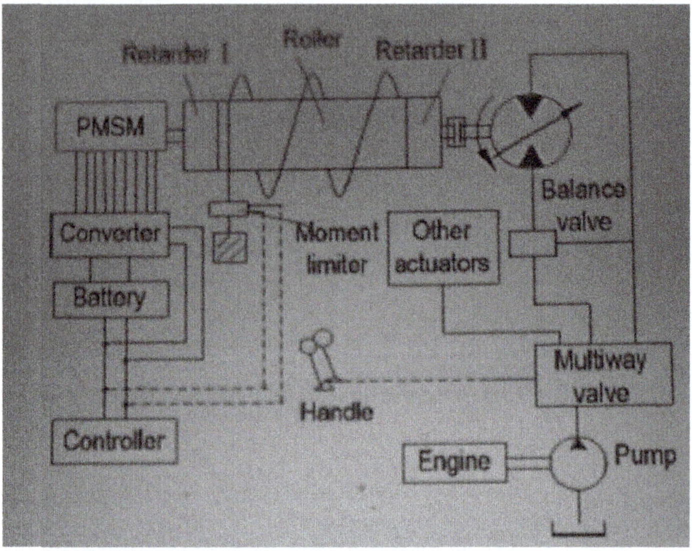

Fig. 3.35 Energy recovery scheme [1]

3.7 Marine

Several on-deck equipment, including cranes, derricks, hoists, and winches, deploy EH servo systems (Fig. 3.36).

Fig. 3.36 On the deck hydraulic equipment [20]

3.7.1 Winches

Anchor and Mooring, Towing, Fishing, and other winches make different demands on the system. These are met with EH systems working at 350 bar pressure in sea water-protected components. They are compact with small footprints [20] (Fig. 3.37).

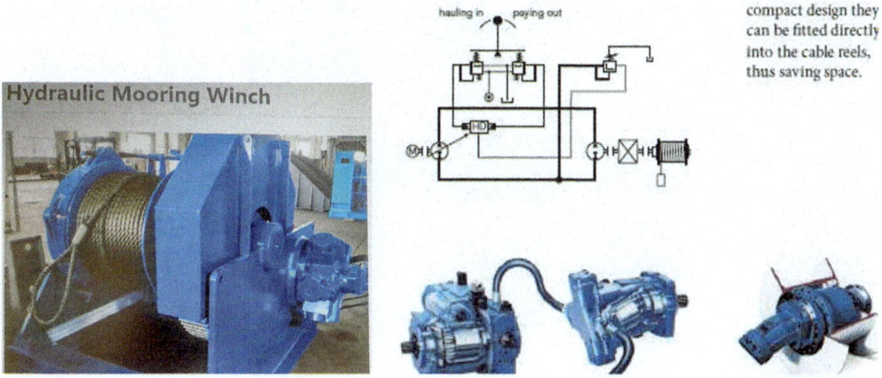

Fig. 3.37 Hydraulic marine winches [20]

3.7.2 Rudder Control

A scheme for rudder control of a ship is shown in Fig. 3.38 [20]. It uses two independent drives fed by separate pumps.

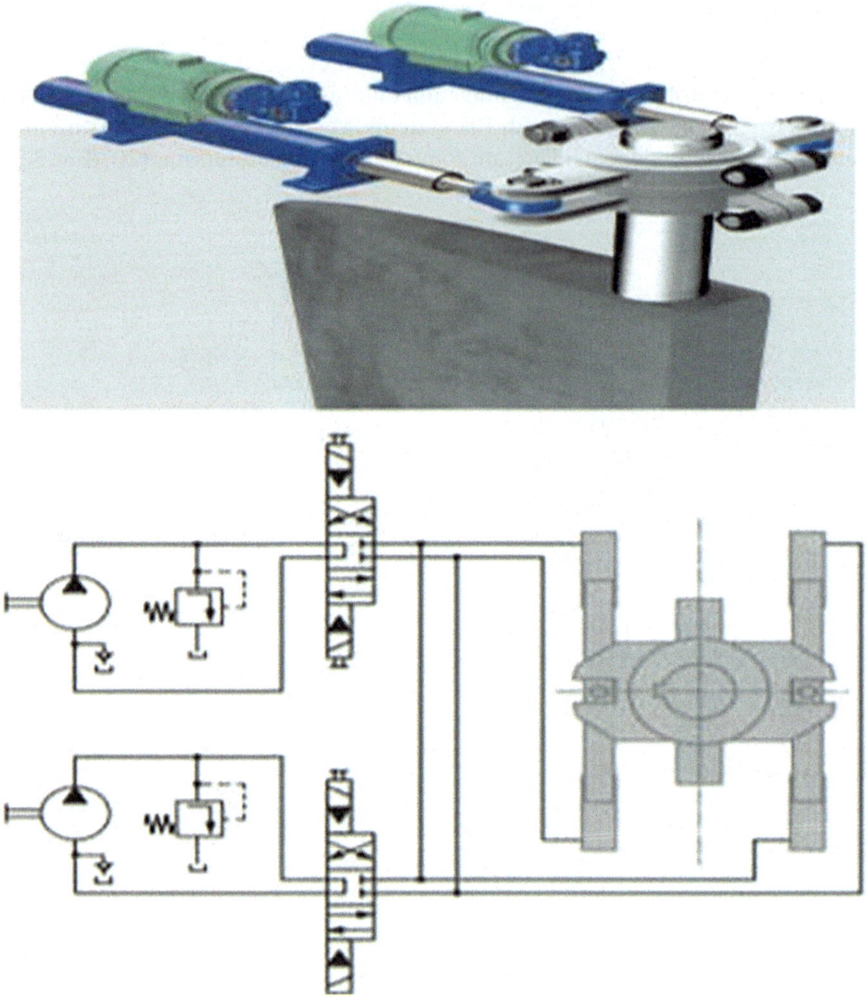

Fig. 3.38 Rudder control [20]

3.7.3 *Ship Stabilizers*

Under rough sea conditions, Fins are extended to the port and starboard sides, and the lift forces on these sides are controlled by the pitch angle. The command is computed by an **acceleration measurement**, which is processed to set the position command [21] (Fig. 3.39).

Fig. 3.39 Ship stabilizer [20]

3.8 Defense

Armored tanks have EH servos for two functions: drive and steering of the tracked vehicle and angular position control of the traverse and elevation with inertial reference.

3.8.1 Track Drive

Two independent servo drive packages that are fed by two pumps control the speed and rotation of tracks. Running tracks in opposite directions at the same speeds can turn the vehicle on its own axis. Independent drives may cause load locking if a decoupling strategy is not adopted [22].

Vehicle speeds up to 70 and 40 kmph on rough terrain are desired. Low-speed high-torque motors working at high pressures are prescribed for application.

3.8.2 Weapon Control System

There are two subsystems: the fire control system and the gun control system.

The FCS, the target generator, sets the reference commands to elevation and azimuth drives of GCS, realized by ballistic computation with sensory inputs, Laser Range finder, IMUs, etc.

The GCS executes these commands with respect to an inertial reference equipped with a gun stabilization subsystem. The need for gun stabilization is apparent from Fig. 3.40 [23].

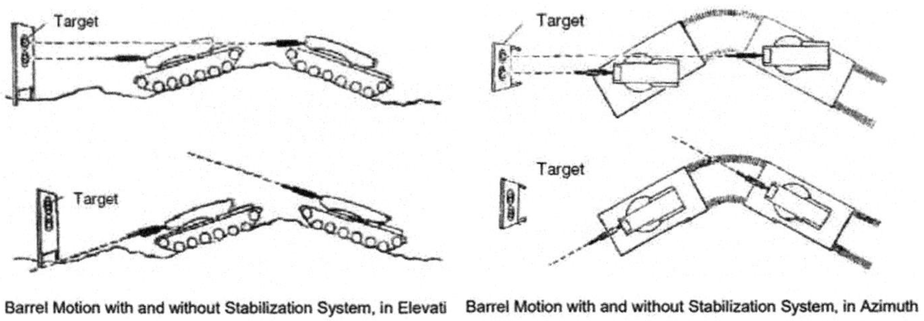

Fig. 3.40 Need for gun stabilization with inertial reference [23]

Gun stabilization in two planes is provided by gyro-based inertial accelerometers, which are processed and fed as corrections to commands to gun control servo systems (Fig. 3.41).

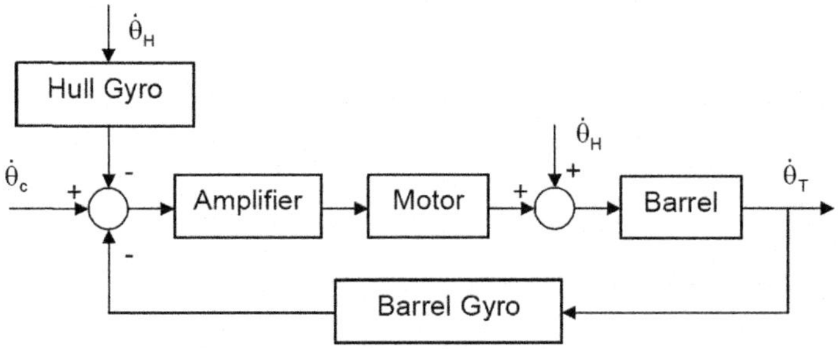

Fig. 3.41 Disturbance compensation [23]

EH servo drive for azimuth and EH servo actuator for elevation control is common. The EH servo actuator provides structural stiffness during recoil. Elevation drives have designs to meet space restrictions and balance heavy-barreled guns.

The azimuth drive utilizes EH servo drive packages.

A challenging requirement of gun control is the disturbance of the gun barrel and hull due to movement and gun recoil.

Disturbance rejection is a candidate for modern control strategies, adaptive and variable structure sliding modes, intelligent control strategies, fuzzy PID, neural networks, and learning systems.

3.8.3 *Missiles*

Missiles are unmanned aircraft that need flying controls for roll, pitch, and yaw movements (Fig. 3.42).

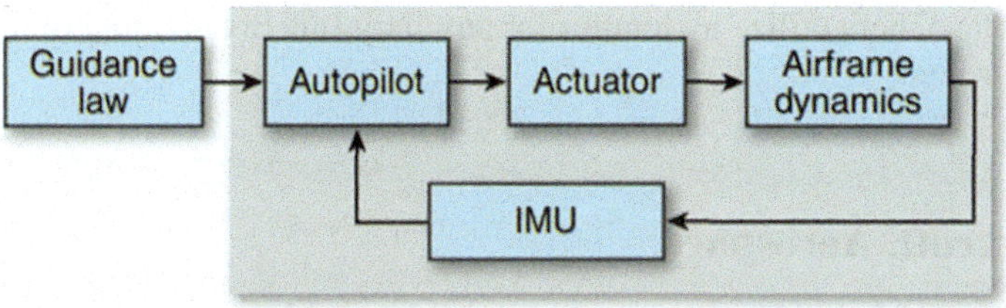

Fig. 3.42 Missile roll control [24]

The IMU senses the inertial movement of the missile. Its outputs and the inputs from guidance law are combined in autopilot to form a command to actuators moving the control surfaces.

Power for drives in missile control is augmented from stored energy in accumulators. To minimize quiescent power loss, the servo valve must minimize its 1-stage leakage flow.

Electrically modulated flow control valves deliver alcohol, liquid oxygen, hydrogen peroxide, and various other fuels and oxidizers to the power element of the missile.

3.8.4 Missile Launchers

Missile launchers deploy EH servo drives for azimuth and EH servo actuators for elevation control. A slewing ring driven by an azimuth drive supports an elevation drive and provides power and data transmission through multiple slip rings. The elevation cylinder, which forms part of a closed triangular structure, adds the desired stiffness for the drive. An elevation drive is shown in Fig. 3.43.

Fig. 3.43 Missile launcher [25]

Missile launchers utilize hydraulics for auxiliary function loading of missiles for a fast turnaround.

3.9 Aircraft, Aerospace

Power to size and power to weight of hydraulic servo systems have made them a drive choice for a number of aircraft/aerospace applications.

3.9.1 Primary Flying Controls

Fly-by-Wire (FBW) control of primary flying control surfaces (Fig. 3.44) is a major application [26].

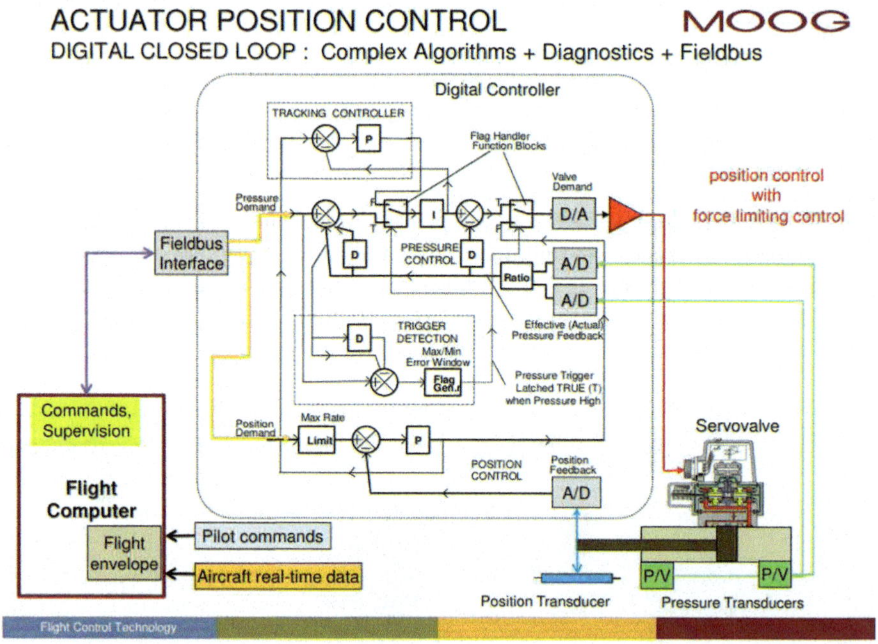

Fig. 3.44 Control loop schematics [26]

The design provides for multiple redundancy in components and temporal redundancy. General schemes are shown in Figs. 3.45 and 3.46.

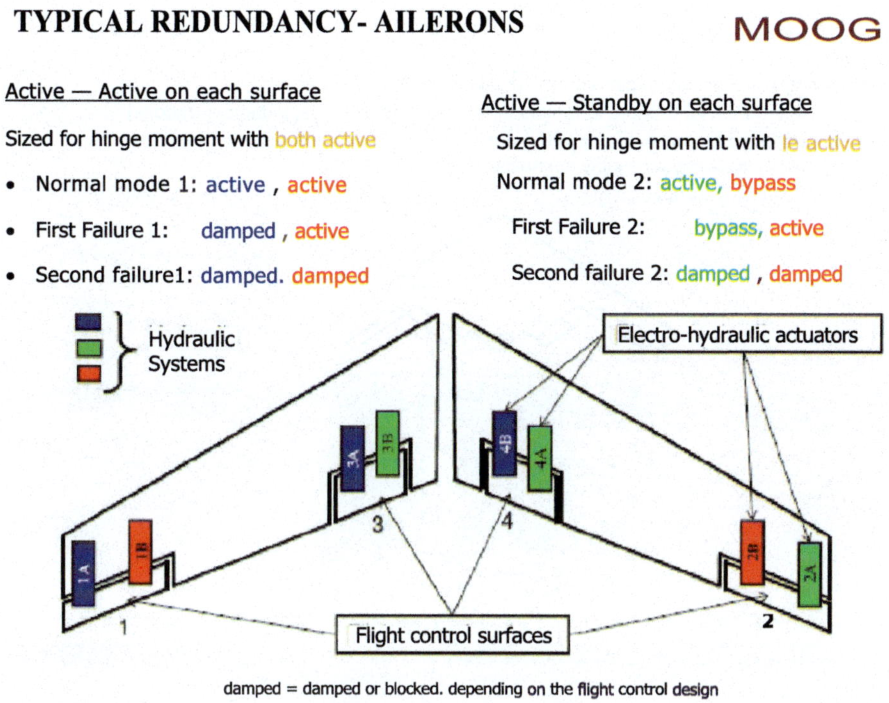

Fig. 3.45 Component redundancy [26]

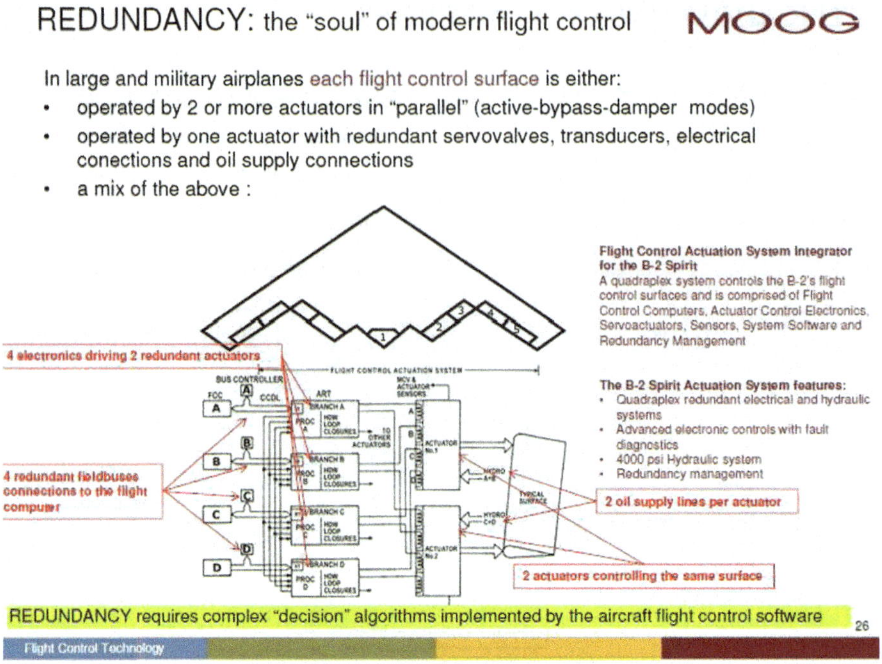

Fig. 3.46 Redundancy in aircraft flight control software [26]

3.9.2 Secondary Flying Controls

Leading edge slats and flaps are EH servo controlled (Fig. 3.47).

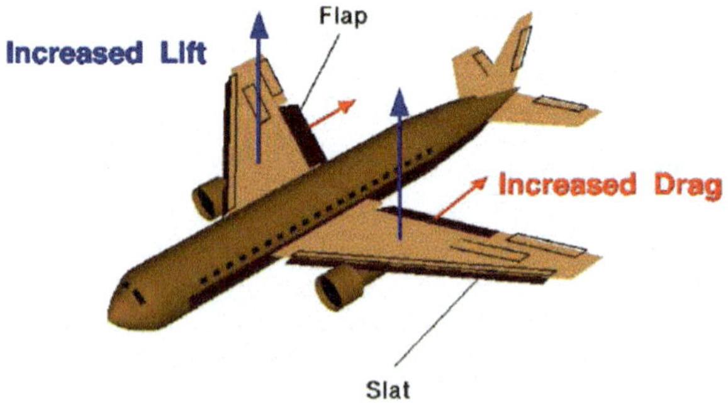

Fig. 3.47 Secondary flying controls [27]

EH servo actuators (EHSA) for flap application triggered the development of these drives for other industrial uses.

3.9.3 Blisk Manufacture

Blisk manufacturing saves considerable weight of engine compressors and turbines. It entails the use of many leading edge manufacturing technologies: linear and rotary friction welding, multi-axis machining, chemical milling, and new additive manufacturing for its repair [28] (Figs. 3.48 and 3.49).

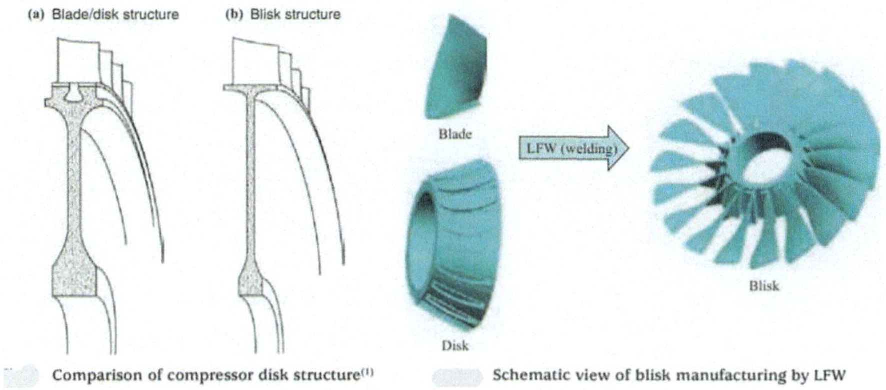

Fig. 3.48 Blisk manufacture

Fig. 3.49 Linear friction welding of Blisk [28]

The LFW machine welds in two planes and carries other functions of loading and orienting parts to be welded. High-frequency relative movement under friction load and subsequent forging demand large controlled flows to the actuators. The system uses 4-stage servo valves.

3.9.4 *Aerospace*

Thrust vector control (TVC) of gimbal-mounted engines in space rockets is controlled by EH argos.

Many of the auxiliary functions in aerospace operations are predicted to be candidates for EHSA application (Fig. 3.50).

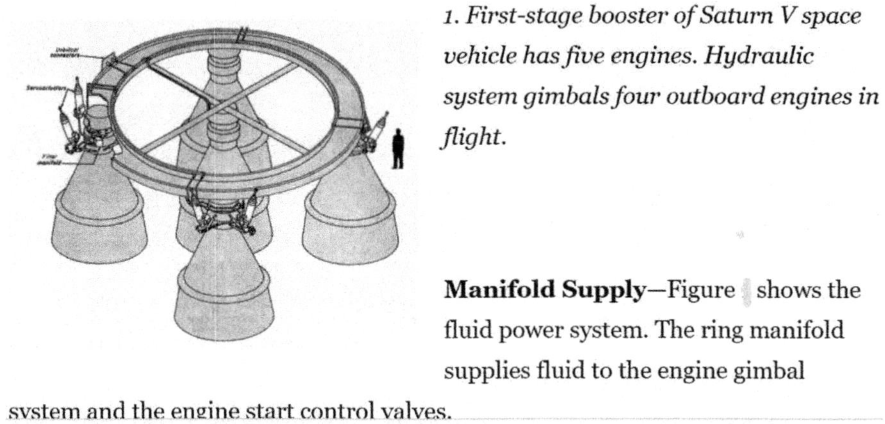

Fig. 3.50 TVC control [29]

Electrically modulated servo valves serve in fuel control of aircraft, aerospace, and marine engines.

3.10 Hydraulic Robots

3.10.1 *Tele Robots*

Robotics technology is propelled by needs of productivity and for hazardous tasks for humans such as the Space, Nuclear, and Subsea domains. Most of these operations are tele-operated, but to alleviate the operator's workload, robots carry out the dynamic function necessary for the task.

3.10.2 *Humanoid Robot*

In developing a humanoid robot, Kawasaki developed a highly integrated EHSA suitable for the legs of KALEIDO (Fig. 3.51).

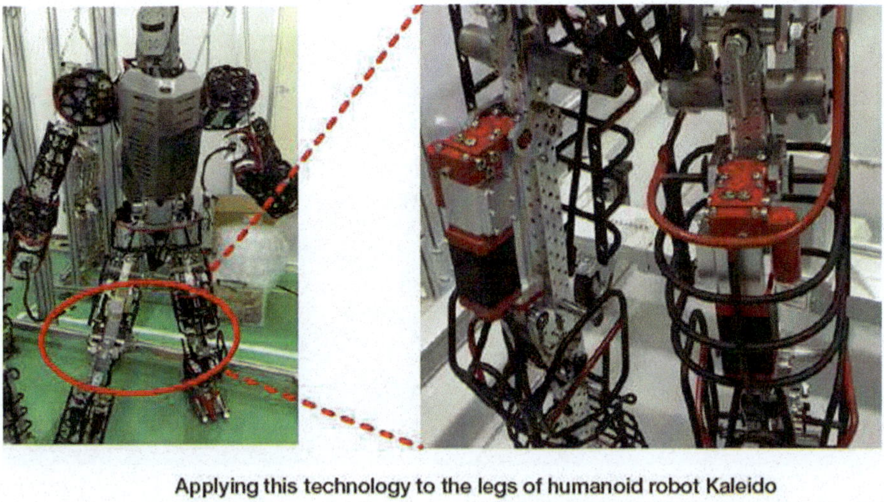

Fig. 3.51 Humanoid robot [29]

3.10.3 HyQ Quadruped Robot

A 3-degree-of-freedom (DOF) bionic leg with 2 rotational movements of the hip and one knee rotation is detailed in [30]. The control of this leg is designed for hybrid position and force control in constrained motion (Fig. 3.52).

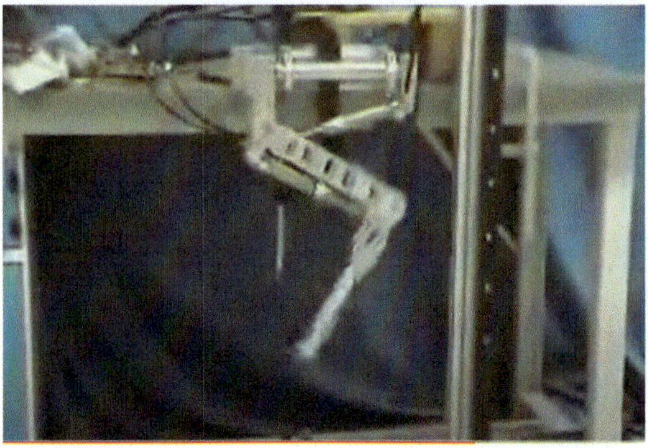

Fig. 3.52 3DOF HyQ leg [30]

Four of these legs are then controlled in a supervisory/hierarchical controller issuing commands to the legs based on the gait of quadruped movements. HyQ quadruped is capable of trotting, jumping, and other bionic movements (Fig. 3.53).

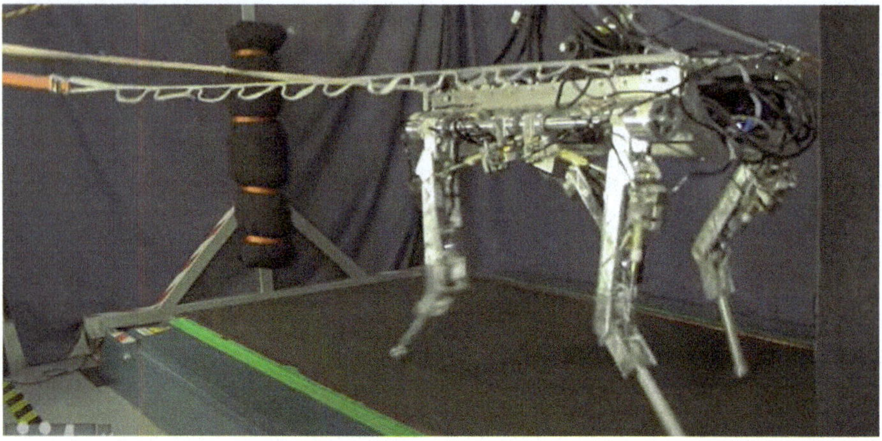

Fig. 3.53 IITs HyQ quadruped robot [30]

3.10.4 The Slingsby TA 9

The Slingsby underwater manipulator is a 6DOF hydraulic robot designed for operations such as weld inspection and docking. The 1.53 m long right-hand robot has 4 joints driven by linear actuators, 2 others by rotary actuators, and a gripper. It is fitted with a 6-axis force sensor for the feedback necessary for hybrid control [31] (Figs. 3.54 and 3.55).

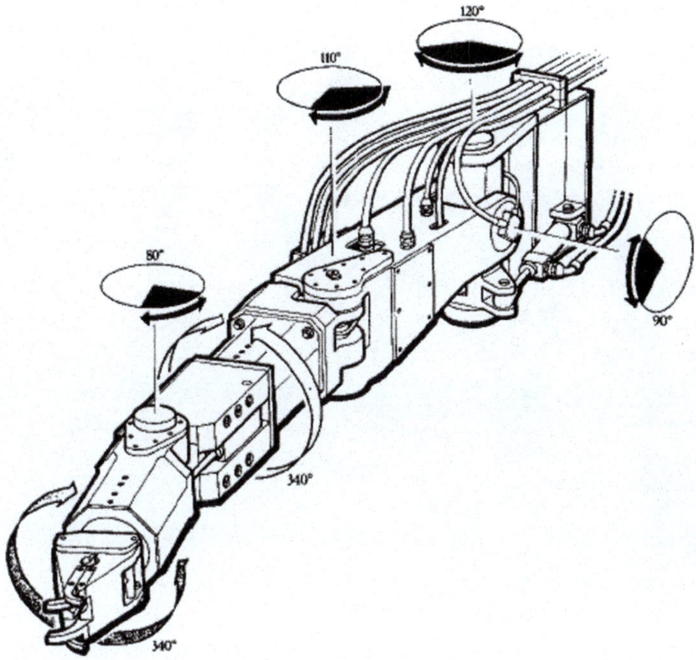

Fig. 3.54 TA9 underwater manipulator [31]

3.10.5 Power of Hydraulic Robots

See Fig. 3.55.

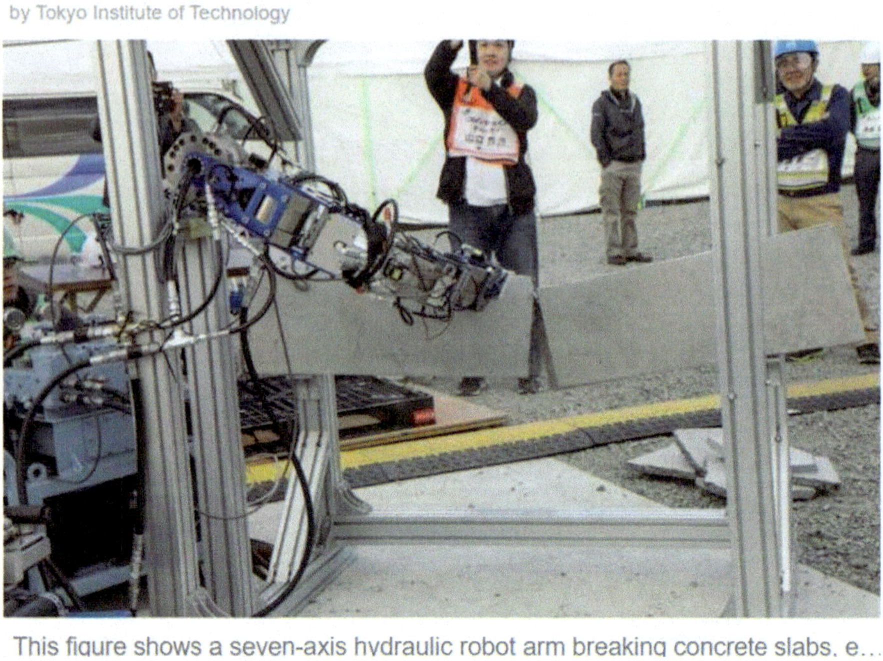

Fig. 3.55 More power to servo hydraulics [32]

3.11 Others

3.11.1 Wind Tunnel Flexible Nozzles

Flexible nozzles vary geometry based on wind tunnel test air speeds (Figs. 3.56 and 3.57).

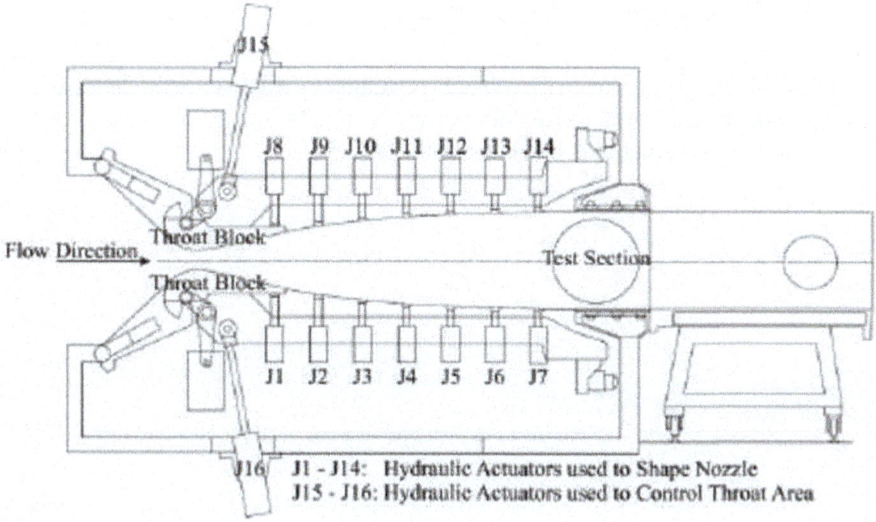

Fig. 3.56 Transonic flexible nozzle [33]

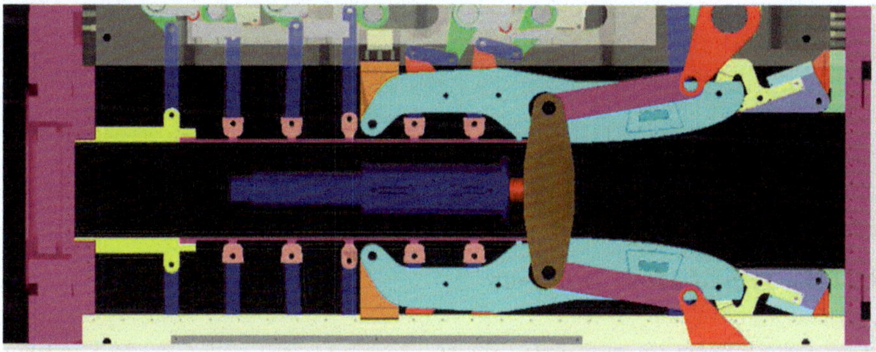

Fig. 3.57 Trisonic VMFN drive system [33]

Earlier designs used multiple EH servo actuators for the purpose of transonic wind tunnels. The discontinuous geometry is now corrected with one actuator driving roller through linkages to provide a smooth contour (spline) of nozzles based on actuator length/tunnel air speed. The drives are powered mainly by stored energy in accumulators, as blowdown is a short-term operation.

As the load is mainly for pushing, the number of accumulators is substantially saved by the use of tandem actuators.

3.11.2 Random Sea Wave Generator

The effect of wave lashes on marine shores is studied by RSWGs. Modeled structures downstream are subjected to random waves by an EH actuator driving, pushing a wave board. Single/dual-axis designs exist (Fig. 3.58).

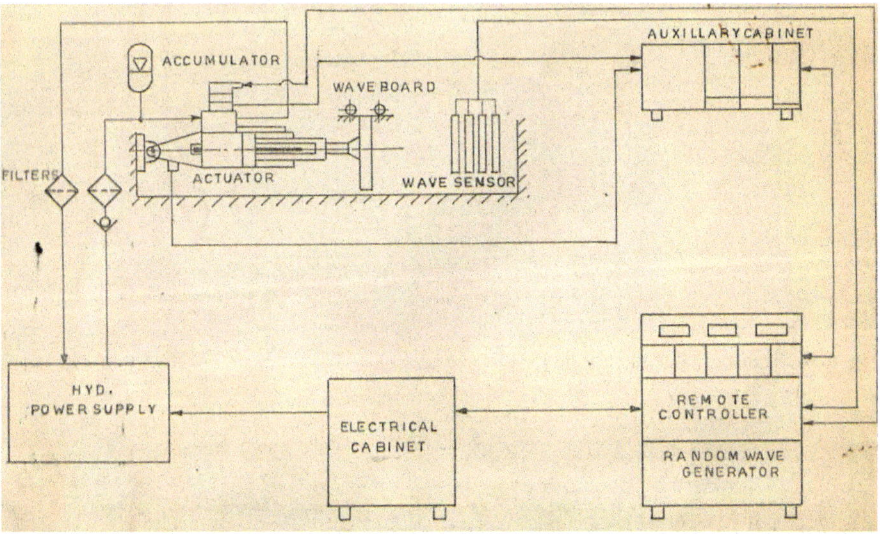

Fig. 3.58 RSWG schematics [34]

3.11.3 Tuned Mass Dampers

Sky scraping structures subjected to wind loads cause low-frequency large-amplitude vibrations, which are unacceptable. Heavy-tuned mass dampers on top of the superstructure are engineered to reduce this amplitude to an acceptable level (Figs. 3.59 and 3.60).

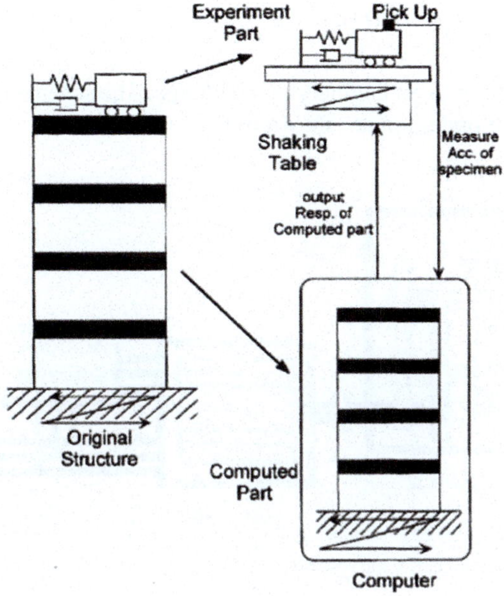

Concept of substructure shake table testing including a tuned mass damper, CREDIT: Igarashi et al. (2000).

Fig. 3.59 Tuned mass dampers on skyscraper [35]

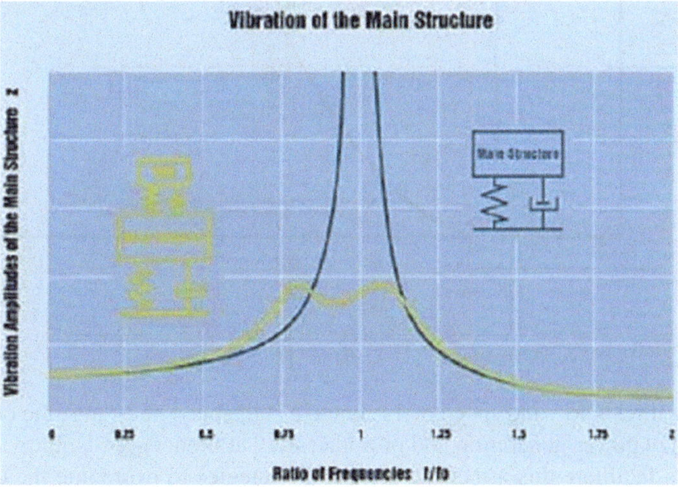

Fig. 3.60 TMD effect [35]. Amplitude—frequency response of a low damped system without (blue) and with (yellow) tuned mass damper

3.11.4 Wind Energy Harvesting

Harvesting wind energy is an engineering challenge. The scheme for harvesting and wind speed range of operation are shown in Figs. 3.61 and 3.62.

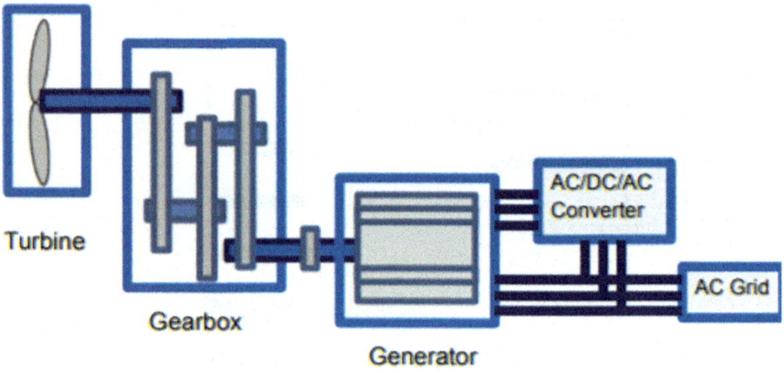

Fig. 3.61 Scheme of wind energy harvesting [36]

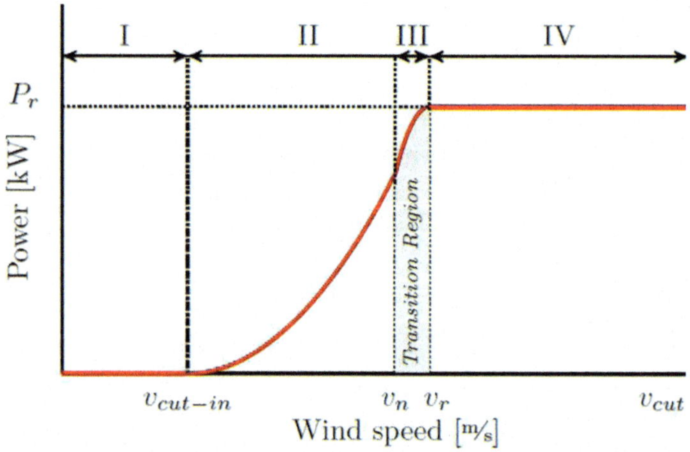

Fig. 3.62 Wind energy harvesting [36]

From a threshold wind speed of 11 m/s to a stall speed of 25 m/s, the objective is to maximize power generation and provide safety at higher speeds. Individual pitch controllers facilitate this armed with control strategies to overcome the challenges of structural disturbances and time delay of the hydraulic system.

3.11.5 ADAS, Theme Parks

Advance Driver Assist Systems (ADAS) and the many Theme Park joyrides are based on 6 DOF motion simulators. The 3 translatory and 3 rotational acceleration simulator inputs are passed through a matrix of "wash out filters", which are switched on cues coming from visual, audio sensors, and/or driver inputs in motion cue algorithms [37] (Fig. 3.63).

Fig. 3.63 DLR driving simulator [37]

3.11.6 Prosthesis Ankle

Reference [38] details the design, control, and testing of an EHSA-powered ankle prosthesis. The prototype ankle prosthesis can operate passively in most of the gait cycle and provide powered assistance to toe push-off and subsequent dorsefication. The prosthesis powered by a 100 W brushless DC (BLDC) servomotor is tested with a 70 kg amputee. Amputee has given positive feedback (Figs. 3.64 and 3.65).

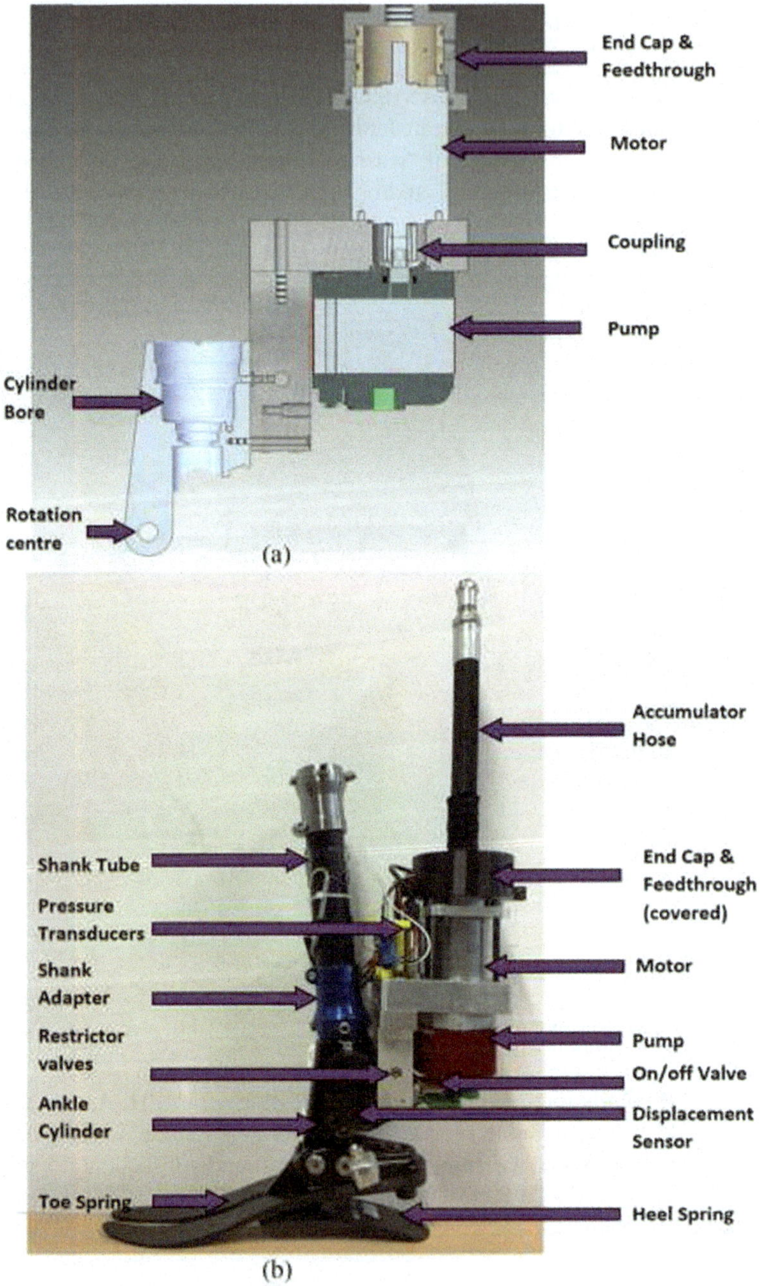

Fig. 3.64 EHSA-powered prosthesis ankle, cross section, prototype [38]

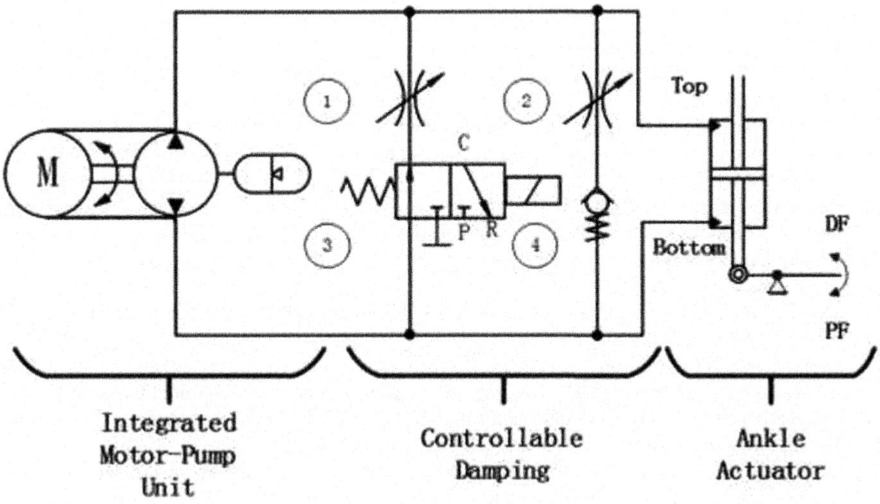

Fig. 3.65 Hydraulic circuit [38]

References

1. ETA Technology (2021) Innovative machines for productivity, performance & profitability. Catalog
2. Rethinking injection molding. Hydraulics-MOOG.moog.com>literature>ICD. Accessed 2 Jan 2022
3. DigiPack Parison Controller. Installation, maintenance and User manual.moog.com>literature. Accessed 24 Apr 2021
4. Prasad SV et al (1994) Development of high speed Globoidal Cam indexing units. In: 16th All India manufacturing technology and research conference, Dec 1994
5. Bert B (1995) On adaptive electrohydrostatic drives.11.IFK. Mar 2018. A....hydrostatic drives. Notes. Linde
6. Fodor S (2017) Toward semiautomation of forestry cranes. Doctoral thesis. UMEA University, Sweden
7. Xaong S et al (2019) Components sizing and performance analysis of hydromechanical power transmission applied to Wheel loader. Energies. MDPI
8. Differential steering. Wikipedia
9. Wang T (2001) Hydraulic power steering system design, optimization, simulation.. In: World conference. SAE, Detroit, Mar 2001
10. Raghavendra DR, Kanavi S (2016) Fluid power technical seminar. Bangalore
11. Cleasby KG (2008) A plummer. A novel high efficiency electrohydrostatic flight simulator motion system. In: Fluid power and motion control. Bath
12. Catalog. Servotest systems. http://www.servotest.com.. 14 Sept 2022 updated
13. Enholm R (2005) Design of landing gear test equipment. Master of science thesis. Helsinki University of Technology
14. Xuan J, Wang S (2018) Development of hydraulically driven fatigue testing of Insulators.IEEE. Accessed Feb 2018
15. Raghavendra DR, Venkataramiah S et al (1977) Performance of electro hydraulic force exciter. In: Annals of CIRP, Delhi

16. Zhao S et al (2004) Effective force testing with nonlinear velocity feedback compensation. In: 13th World conference on earthquake engineering, Canada
17. lianpeng Z et al (2015) Modeling and vibration decoupling of multi axial shaking table. In: International informatic computer engineering conference. IIIE
18. Raghavendra DR (1988) Hydraulic systems Handbook. Utility Publications Ltd.
19. Chen Y et al (2018) Investigation of energy regeneration and control strategy of crane hoisting system. J Mech Eng
20. RexRoth Bosch Catalog. Hydraulic system solutions in marine technology. Re09907.dede.resource.bosch.com>media>marine. Accessed 20 Apr 2021
21. Anish. Understanding ship fi stabilizers and its operation. http://www.marineinsight.com. Accessed 07 May 2021
22. Zhu C, Zhang H, Wang L, Li K, Liu W (2020) Robust control of hydraulic tracked vehicle driven system with quantitative feedback theory. Int J Distrib Sensor Netw 16(2) (SAGE)
23. Gumusay O (2006) Intelligent gun stabilization control of Turret subsystem under disturbances from unstructured terrain. Thesis, Master of science. Middle East Technical University
24. Chapter 5 missile control system. SanFrancisco Maritime National Park Association. Catalog
25. Missile launcher hydraulic systems. York Precision. Catalog
26. Valdo M (2012) Servo hydraulic technology in flight control. In: Workshop on innovative engineering for fluid power and vehicular systems. Sao Paulo
27. Flaps and slats-NASA. www.gic.nasa.gov
28. Adams D (2019) Whiteboard Wednesday: linear friction welding of Blisk. Manufacturing Technology Inc. 2 Oct 2019:9:01am
29. Hadel J, Church RF (2017) Hydraulic controls for gimballing Saturn V engines. Hydraulics and Pneumatics
30. Focchi M et al (2010) Control of hydraulically operated quadruped robot leg. In: International conference on robotics and automation, Alaska, US
31. Clegg AC (2000) Self tuning position and force control of a hydraulic manipulator. Doctoral thesis. Heriot Watt University
32. New hydraulic actuator will make robots tougher-EurekAlert. http://www.eurekalert.org/news-releases. Accessed 17 Sept 2022 updated
33. Ramaswamy MA (2013) Invited to talk on VMFN. In: 3rd International conference on wind tunnel testing, Trivandrum, Aug 2013
34. Jones P, Raghavendra DR (1990) Performance of a sea wave generator. In: 6th All India exhibition and conference, Pune, Jan 1990
35. Igarashi A et al (2004) Experimental simulation of coupled response of a structural system using substructure resting methods. In: 13th world conference on earthquake engineering, Vancouver, Canada, Aug 2004
36. Gao R, Gao Z (2016) Pitch control of wind turbine system using optimization estimation and compensation. Renew Energy J
37. Weib C (2006) Control of a dynamic driving simulator: time variant motion cueing algorithms and prepositioning. Diploma thesis. Institut fur Verkehrsführung und Fahrzeugsteuerung
38. Yu T, Plummer AR, Iravani P, Bhatti J, Zahedi S, Moser D (2019) The design, control and testing of an integrated electrohydrostatic powered ankle prosthesis. IEEE/ASME Trans Mechatron 24(3)

Chapter 4
Design of SISO EH Servo Systems

Abstract The various Steps in the design process are detailed with respect to a case study.The derivation of Load locus to size power elements, A heuristic approach to estimate closed loop performance from this are presented. It includes Bench Testing means to parameter estimations necessary for grey box models used in Control design.

4.1 Valve-Controlled Position Servo

The design of a servo valve-based position-controlled system is discussed here with reference to a case study. The design entails sizing of all the power elements hydraulic actuator, servo valve, and hydraulic power supply along with the feedback transducer. The design is initiated by a set of coherent specifications in terms of load and velocity over a range of operations together with system requirements of accuracy and response. An optimized kinematic arrangement of actuator with respect to load is a necessary input to design.

The design of an elevation drive of a rocket launcher is described as a case study. The aim is to present the range of calculations involved in sizing the power elements and a heuristic approach to estimate system performance.

The kinematic arrangements for the elevation drive are shown in Figs. 4.1 and 4.2.

D. R. Raghavendra, *Electrohydraulic Servo Systems*, Springer Tracts in Mechanical Engineering, https://doi.org/10.1007/978-981-19-8065-7_4

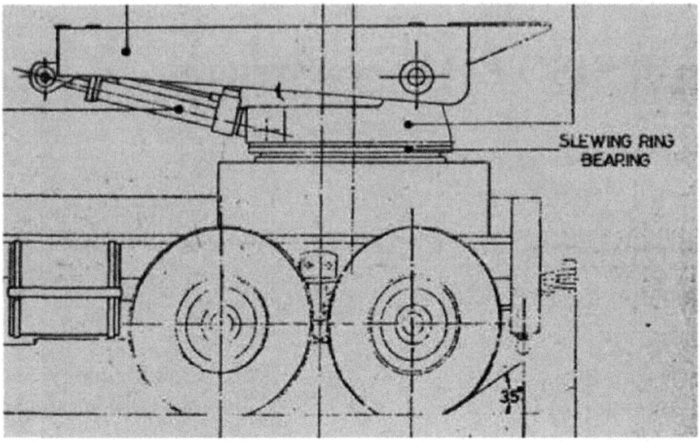

Fig. 4.1 General arrangement

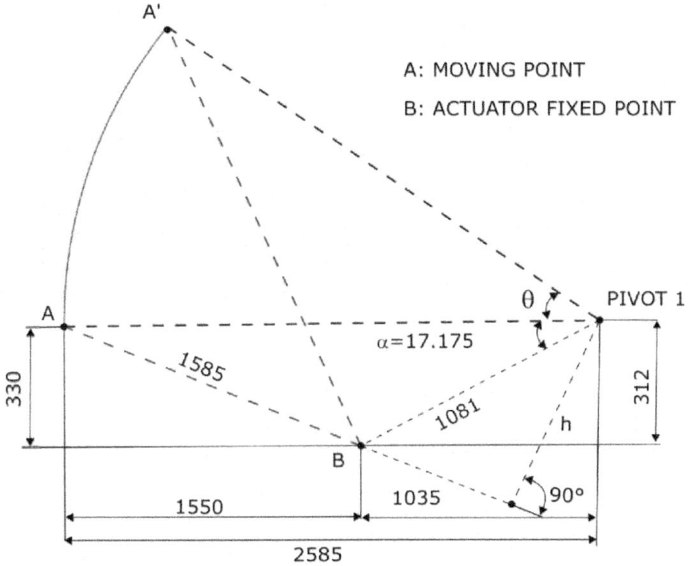

Fig. 4.2 Kinematic arrangement

4.1.1 Main Specifications

Moving Mass	5300 kg
CG	0.815 m from pivot

Inertia	12500 kgm^2
Range of laying	0–55°
Rate of laying	5°/s
Accuracy (dead reckoning)	<1 mrad

4.1.2 Design Methodology

The practiced methodology is to derive the engineering specifications from the general specifications in terms of load locus—force and velocity over the working range of operation. The power elements (actuator and servo valve) are sized to meet the load locus requirement with adequate margin. A preliminary estimate of accuracy and response is then made aided by heuristics (Fig. 4.3).

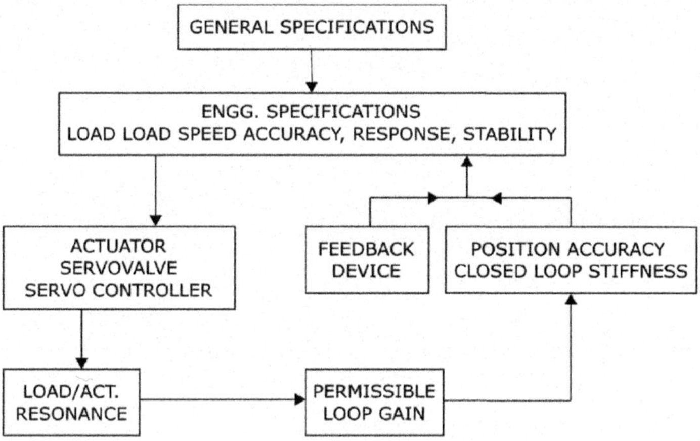

Fig. 4.3 Design methodology

4.1.3 Load Locus

From kinematics,

Cyl. Center distance (in mm)

$$r = 1000\sqrt{7 \cdot 85 - 5 \cdot 59 \cos(\theta + 17 \cdot 17)} \tag{4.1}$$

The stroke requirement is derived from this. (1000 mm)

$$\frac{\mathrm{d}r}{\mathrm{d}t} = \frac{559 \times \sin(\theta + 17.17) \times \omega}{2r} \tag{4.2}$$

The cylinder velocity is derived by differentiating Eq. 4.1 at a specified maximum laying rate of $5^{\circ}/\sec(\omega)$ (Fig. 4.4).

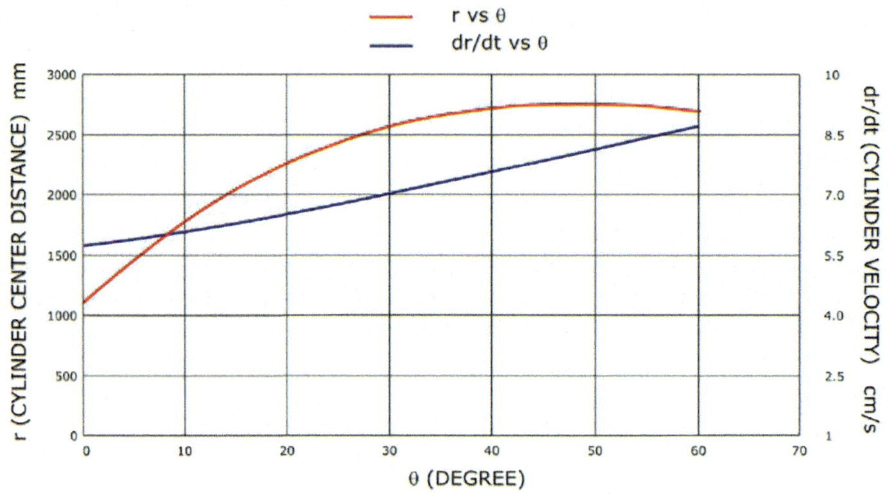

Fig. 4.4 Cyl center dist, cyl. velocity versus θ

Torque on pivot versus θ is calculated from

$$T = 5300 \times 0.815 \times \cos(\theta + 20.5) \tag{4.3}$$

Perpendicular distance h, from pivot is from equation

$$h = 2585 \times 1081 \cdot 4 \times \frac{\sin(\theta + 17 \cdot 17)}{r} \tag{4.4}$$

See Figs. 4.5 and 4.6.

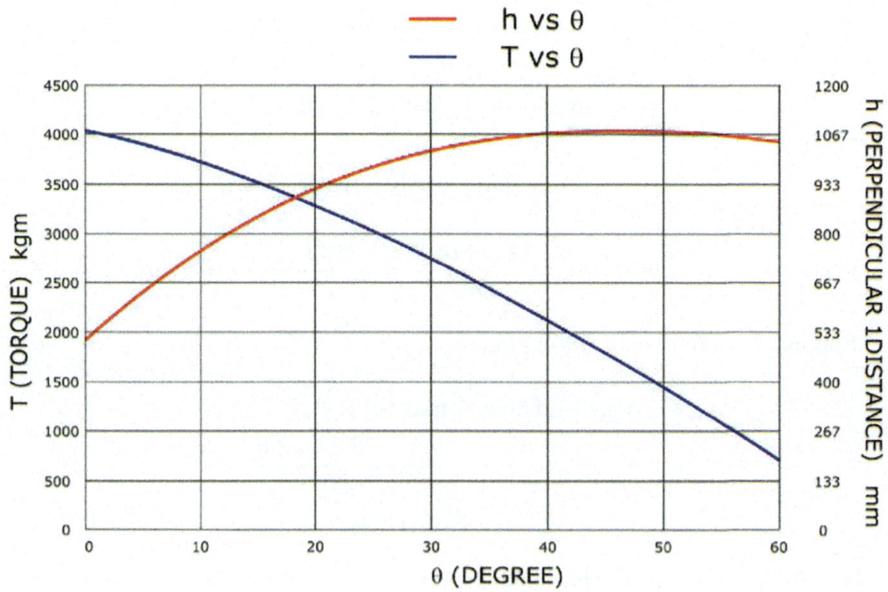

Fig. 4.5 Torque and perpendicular distance versus angular position

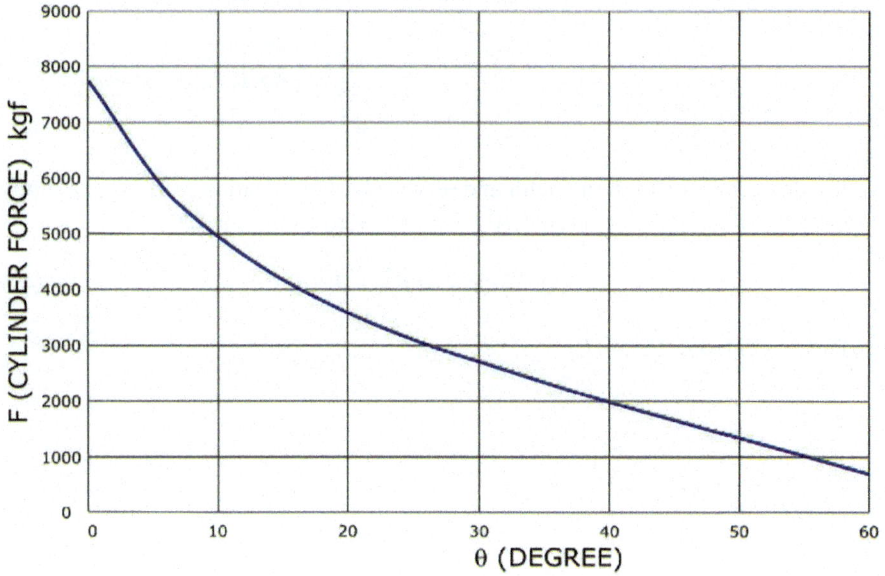

Fig. 4.6 Force versus angular position

Note that the maximum load on the cylinder is 7800 kgf, and a standard cylinder with a 100 mm bore, 56 mm rod, and 1000 mm stroke is designed.

$$\text{Cyl Head endarea } (A_h) = 78.54\,\text{cm}^2$$

$$\text{Annular area (Ar)} = 53.9\,\text{cm}^2$$

$$\text{Area Ratio} = \text{AR} = 1.45$$

$$\text{max load pressure} = \frac{\text{Max.Load}}{\text{area}} = \frac{7800}{78 \cdot 54} = 100\,\text{kgf/cm}^2$$

Supply pressure is set at 120 kgf/cm^2

$$\begin{aligned}\text{max load flow} &= \text{max vel} \times A_h \\ &= \frac{9.4 \times 78.54 \times 60}{1000} \\ &= 44.3\,\text{lpm}\end{aligned}$$

Say, 48 lpm including valve leakage flow.
The servo valve rated at 57 lpm at a valve pressure drop of 70 kgf/cm^2 is selected.

$$\begin{aligned}\text{Valve Pressure Drop Pv} &= 70 \times \left(\frac{\text{Ql}}{\text{Qr}}\right)^2 \\ &= 70 \times \left(\frac{48}{57}\right)^2 \\ &= 45\end{aligned}$$

The drive capacity of the actuator and servo valve combination working at a supply pressure of 120 kgf/cm^2 at a rated 100 % input is given by Eq. 4.5.

$$\text{Ql} = \text{Qr} \times \sqrt{Ps - Pl} \tag{4.5}$$

This is plotted in Fig. 4.7.

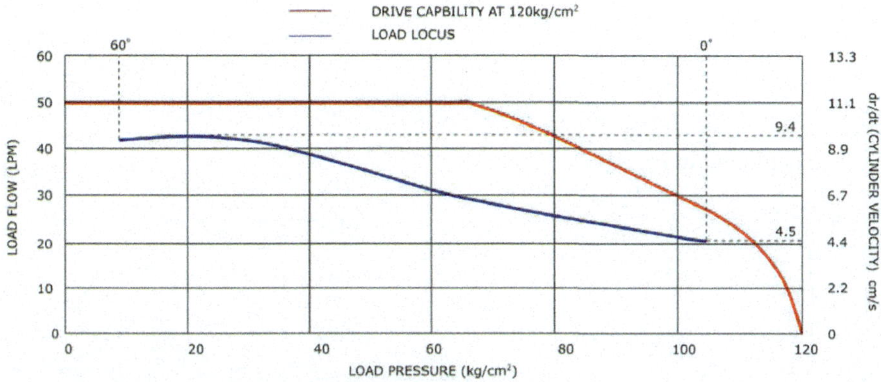

Fig. 4.7 Drive capability versus drive requirement

A power supply of 50 lpm at a pressure of 120 kgf/cm^2 meets the requirement. A pressure-compensated variable displacement pump is recommended based on efficiency considerations.

4.1.4 System Performance Estimation

Neal [1] provides a heuristic approach to estimate system accuracy and response from this combination of actuator/servo valve/load. The system accuracy, response, and stability are critically dependent on the loop gain of the system, which is limited here by the load/actuator resonance. From a stability consideration, loop gain K $< 2\delta_h\omega_h$ [2]. Load actuator resonance (ω_h) is a function of hydraulic stiffness and the inertia seen by the cylinder.

This is evaluated in the following steps:

- hydraulic stiffness versus cylinder position,
- hydraulic stiffness versus elevation angle and
- inertia seen by actuator.

The hydraulic drive system is a function of trapped oil volumes and the bulk modulus of oil and varies with position along the stroke.

$$K_h = \beta_e \times \left\{ \frac{A_h^2}{V1 + A_h \times l} + \frac{A_r^2}{v2 + \text{Ar} \times (S - l)} \right\} \tag{4.1}$$

where $\beta_e = 10,000\,\text{kgf/cm}^2$(recommended value)

$$v1 = v2 = 500\text{cc (assumed)}$$

$$1 = \text{piston displacement}$$

$$S = \text{Total stroke}$$

This is plotted in Fig. 4.8

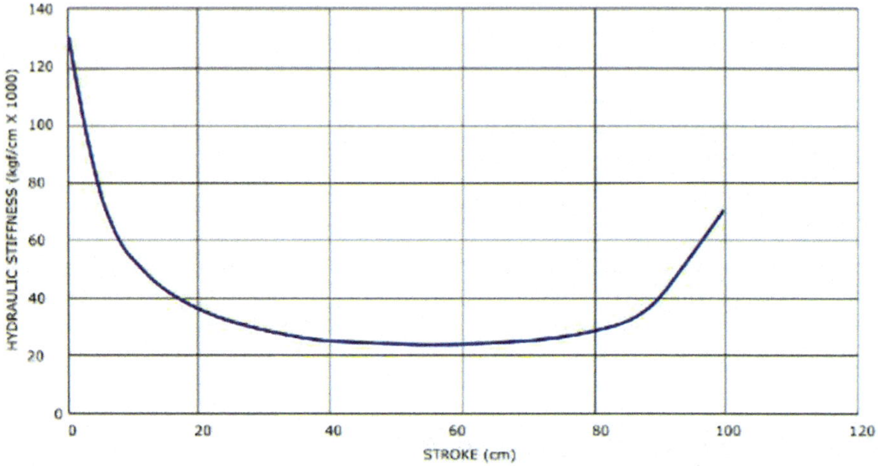

Fig. 4.8 Hydraulic stiffness versus cylinder position

Knowing the relation between stroke and elevation angle, Fig. 4.9 plots Kh versus θ.

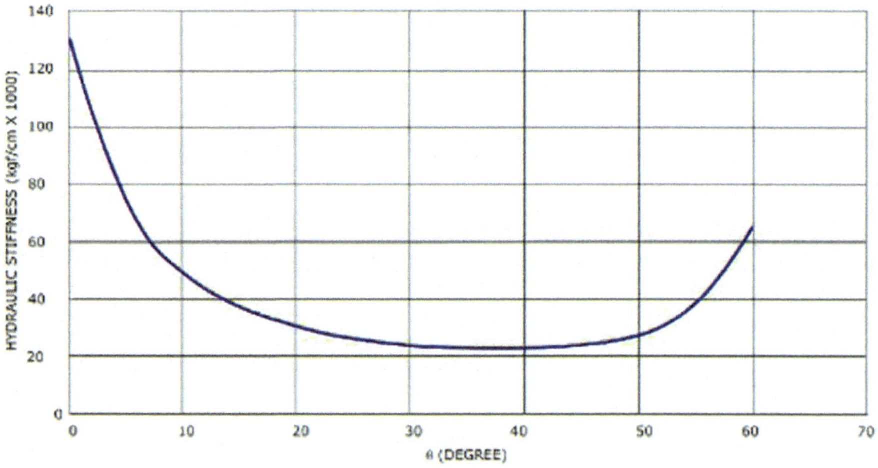

Fig. 4.9 Hydraulic stiffness versus elevation angle θ

Inertia seen by the cylinder is derived by equating energies.

$$\frac{1}{2}\mathrm{M}v^2 = \frac{1}{2}\mathrm{I}\omega^2 \tag{4.2}$$

The relation between the elevation angle and cylinder velocity is shown in Figs. 4.10 and 4.11.

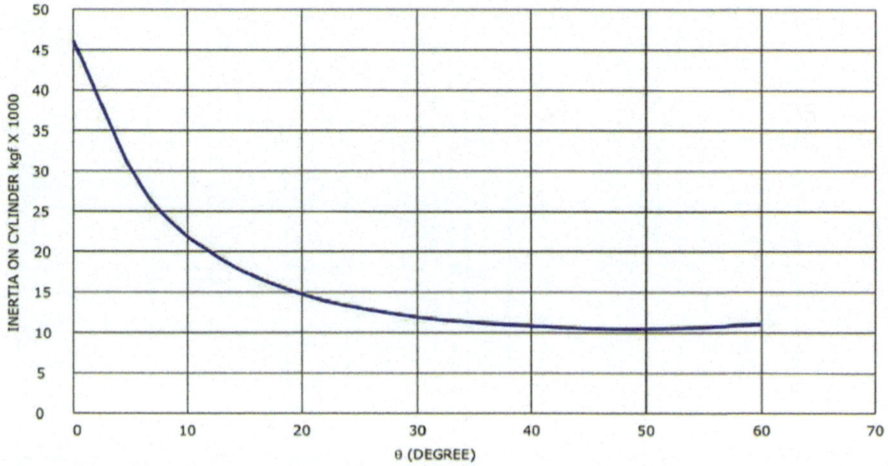

Fig. 4.10 Inertia seen by cylinder versuselevation angle

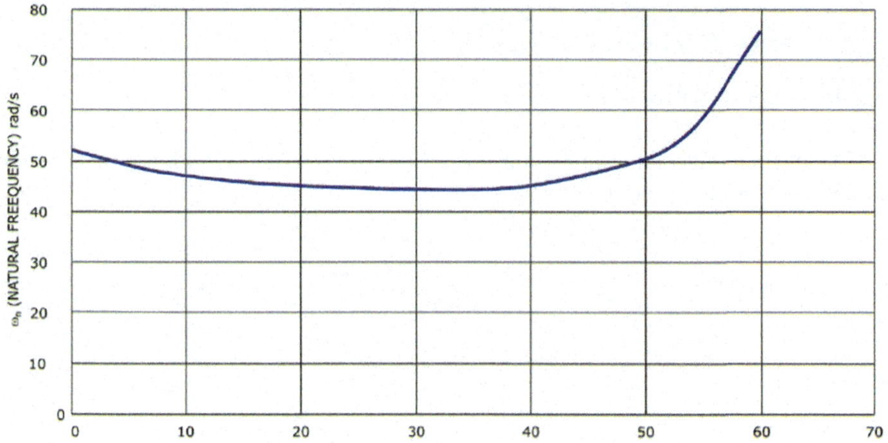

Fig. 4.11 Hydraulic natural frequency versus angular position

$$\text{Hydraulic natural frequency} = \sqrt{\frac{\text{Hydraulic stiffness}}{\text{inertia seen}}}$$

From the above, the minimum natural frequency = 44.2 rad/s.

Hydraulic systems are poorly damped. MOOG, who are pioneers in EH Servo technology, recommends permissible loop gain ensuring a minimum stability gain margin of 6 dB to be 0.1 times the natural frequency [1] (Fig. 4.12)

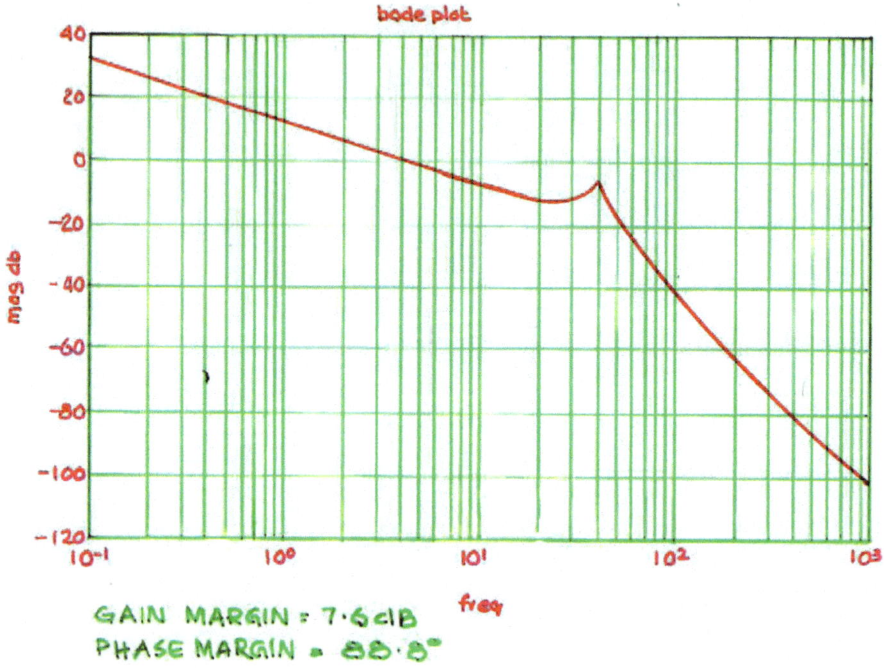

Fig. 4.12 Open loop Bode plot

The expected closed loop response is a first-order response with a time constant of 250 ms (1/k) [1] (Fig. 4.13)

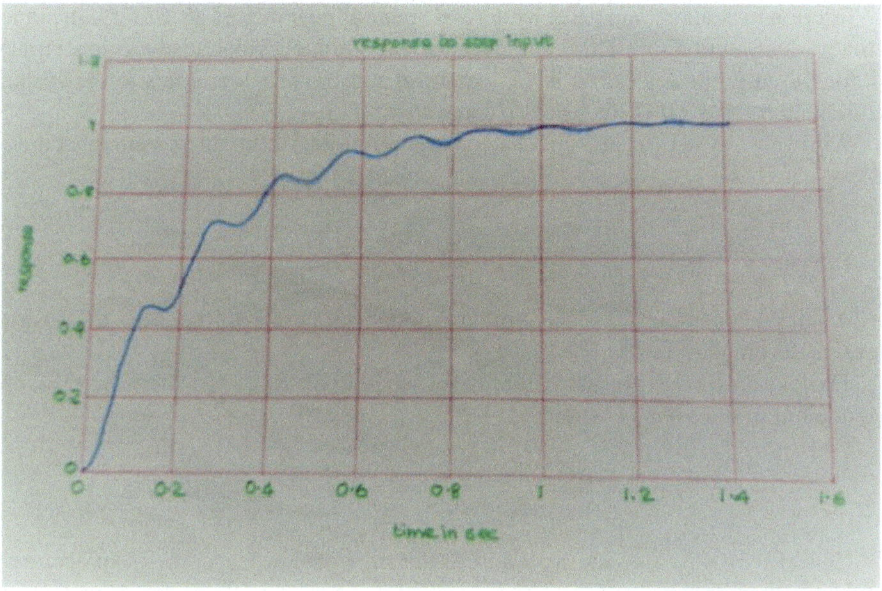

Fig. 4.13 Step response

The normal inputs to systems are higher-order polynomials with a trapezoidal velocity profile or the *S* curve as used in robot motion trajectory planning software. This ensures that the output position reaches the final destination at the same instant. This enables suitable subsequent logic operations (Fig. 4.14)

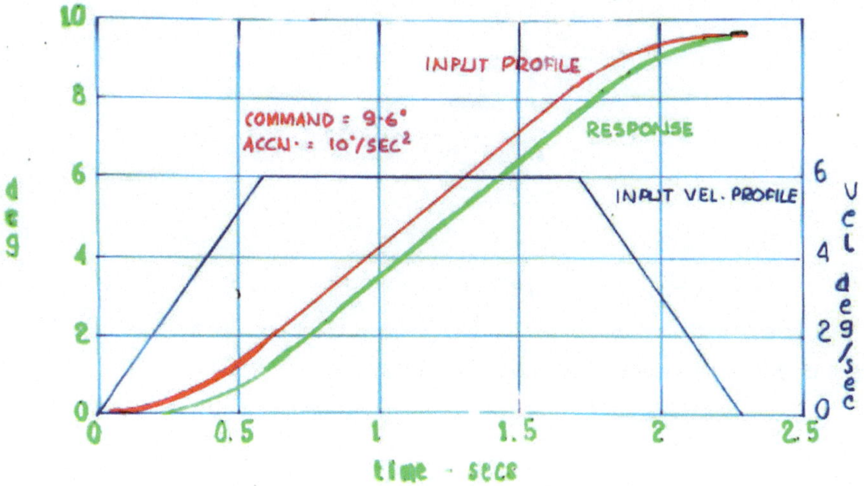

Fig. 4.14 Motion trajectory with trapezoidal velocity profile

Friction in actuator and valve hysteresis are nonlinear factors affecting static inaccuracy (dead reckoning). This system inaccuracy summed with transducer inaccuracy is compared with the desired level. The level sensors provide inputs to the system command for the errors in elevation caused by them.

With valve hysteresis at 3 and 1% for overcoming friction, the system inaccuracy is estimated as

$$\text{Inaccuracy} = \frac{\text{Max.speed}}{25 \times k} = \frac{5 \times 0.04 \times 60}{4} = 3 \text{ arc min}$$

A suitable position transducer with a minimum accuracy of ten times the desired level is designed.

The closed loop stiffness of the drive depends on Kce, and the flow pressure coefficient of the drive is estimated [1]

$$\frac{F}{X} = \frac{20 \times k \times \text{stallforce}\; 20 \times 4 \times 9600}{\text{max vel} \quad 9.4} = 81,700\,\text{kgf/cm}.$$

Closed loop stiffness and load add to the final design inaccuracy. An accuracy budget is then drawn.

If any of the specifications are not met or unsatisfactory, the design is iterated with a new set of power elements or kinematics Figs. 4.15, and 4.16

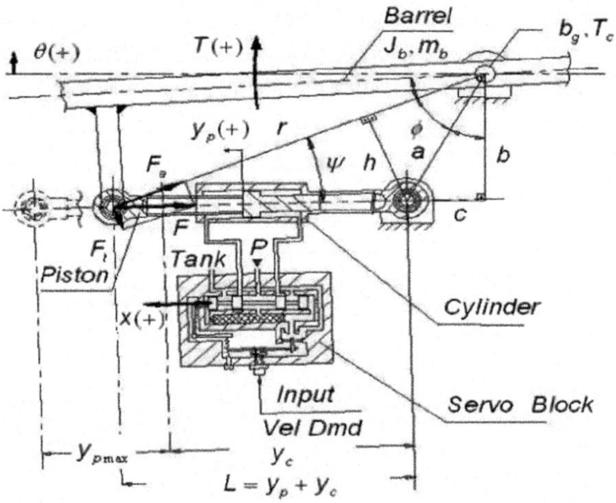

Fig. 4.15 Alternate kinematics for elevation control [3]

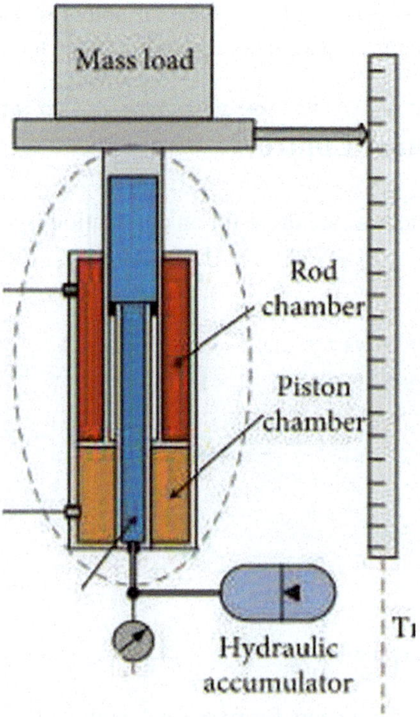

Fig. 4.16 Alternate design using 3-chamber tandem actuator

Alternate design employs a 3-chamber actuator, which provides effective counterbalancing with a separate pump and a proportional relief valve. Accumulators are not used in such applications.

4.1.5 Methods to Improve the Damping Factor

It is apparent from the above that higher damping factors are desirable. The suggested methods to improve damping are as follows:

- Leakage across actuator by a metering orifice. Not recommended as it reduces drive stiffness and increase in damping is minimal.
- Leakage across the valve. Not acceptable as it reduces drive stiffness and increase in damping is minimal increase.
- Load pressure feedback implemented by measuring the actuator chamber pressures. The damping factor increases at the expense of drive stiffness.
- This is recommended when a high-pass filter at 1/3 of the load actuator resonance frequency is included.

- Acceleration and velocity feedback (state variables) is the most desirable method of increasing the damping factor, as it also improves the drive stiffness.

4.2 Rotary Position Control

A block diagram of the traverse drive for an application is shown in Fig. 4.17.

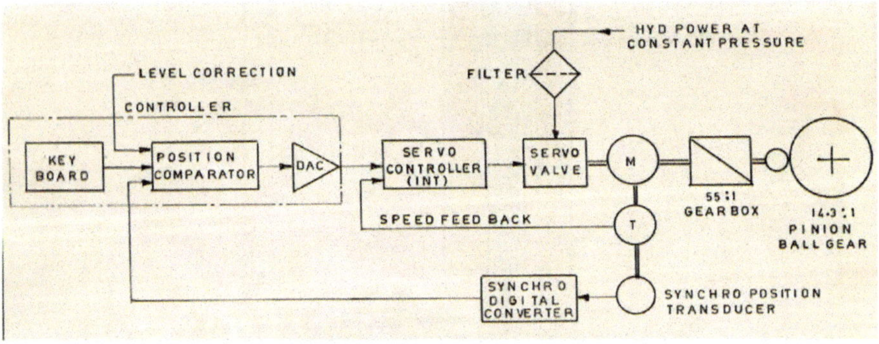

Fig. 4.17 Block diagram of traverse drive

In EH servo drives using gears for rotational angle/velocity control, the selection of gear ratio must be based on the desired load inertia to be met as well as the motor inertia. Load acceleration is critically dependent on this ratio $\frac{\mathrm{Il}}{\mathrm{Im}}$ (Fig. 4.18).

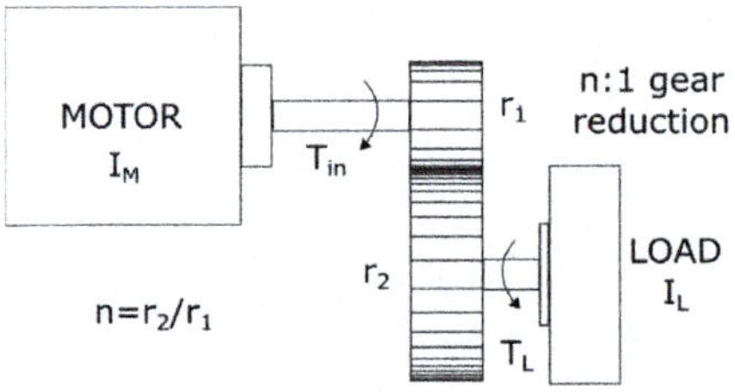

Fig. 4.18 Model of geared motor drive

$$\text{Load Acceleration} = \frac{\mathrm{Tm}}{\mathrm{Il}/\mathrm{n} + \mathrm{Im} \times \mathrm{n}}$$

Nondimensional plots of load acceleration versus N for Il/Im as parameter points to sensitivity (Fig. 4.19).

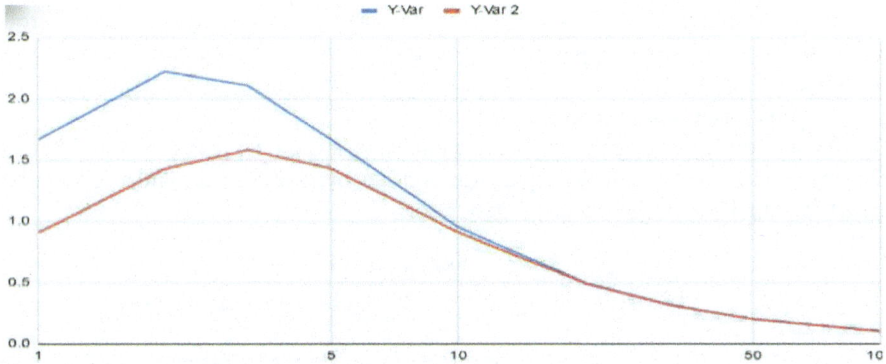

Fig. 4.19 Sensitivity of load acceleration to gear ratio

4.3 Force Control Systems

Force control systems are sought for structural and material testing for strength and endurance. When the load requirement is mainly tension as in fatigue testing or in compression as in vibration testers, tandem actuators are designed. Systems are rated for tonnage at 90 % of supply pressure. Flow requirements are low, as small compressibility flow provides for pressure build-up, and displacement flow is low. The hydraulic power supply should be sufficient to meet the force and velocity demands of application.

Force control systems with proportional controllers are subject to finite errors due to leakage flows. PI control eliminates this steady-state error and meets the needs of most force control systems [4]. Peak valley control by modulation of system input is often resorted. Advanced control schemes are necessary where forces are required around the resonant frequency of the structure, as in seismic testing, as they are subjected to "natural velocity feedback".

The permissible loop gain is limited by the servo valve response and is set at 0.4 times the servo valve natural frequency [2]. The closed loop system response with this setting is 0.5 times the servo valve natural frequency with a damping ratio of 0.5 [2]

4.4 Bench Testing

The preliminary design identifies power elements. The servo valve characteristics, nominal coefficients of flow gain kq, pressure gain kp, and flow/pressure coefficient kce need to be identified, as the catalog values are approximate. Flow gain variation of $\pm 10\%$, kp as 30 to 80% of supply pressure for 1% valve input are quoted.

Control strategies are based on models that can be *white boxes*—derived from actual physical relations of parameters, or *gray boxes* where parameters are tested on bench rigs and identified.

Gray box models are preferred.

The methods of testing the servo drive packages are shown in Fig. 4.20 to identify the system parameters. The drive package consists of a servo motor and a servo valve coupled with a backlash-free tacho generator.

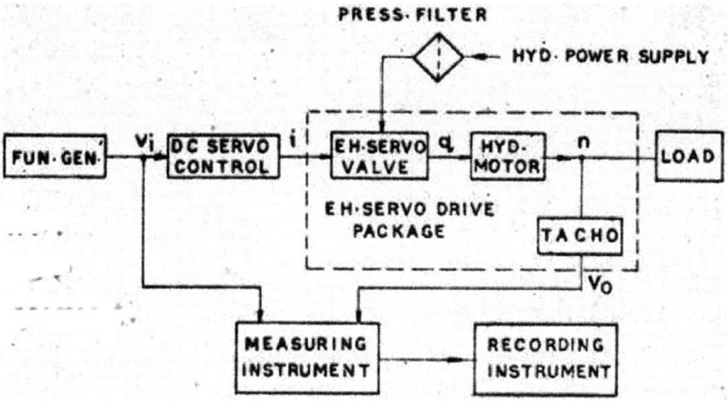

Fig. 4.20 Servo drive test setup

A function generator, differential pressure transducers, and a frequency response analyzer complete the setup (Figs. 4.20 and 4.21).

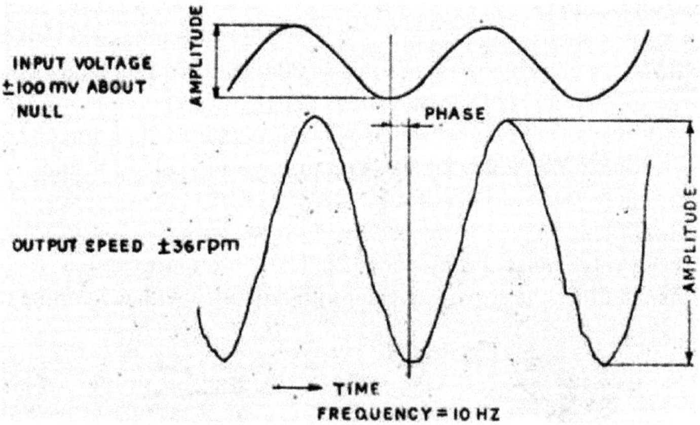

Fig. 4.21 Output waveform distortion

Harmonics in the output speed waveform are eliminated, and flow gain measurements taken at different supply pressures for a range of valve inputs yield no load flow gain, Pl versus Ql data.

The Blocked Rotor Test provides pressure gain of the package.

System dynamic characteristics are measured under simulated inertial loads by subjecting the system to standard inputs.

Sine Sweep across the frequency range of interest picks up the frequency response of the UUT loaded with the simulated initial load (Fig. 4.22).

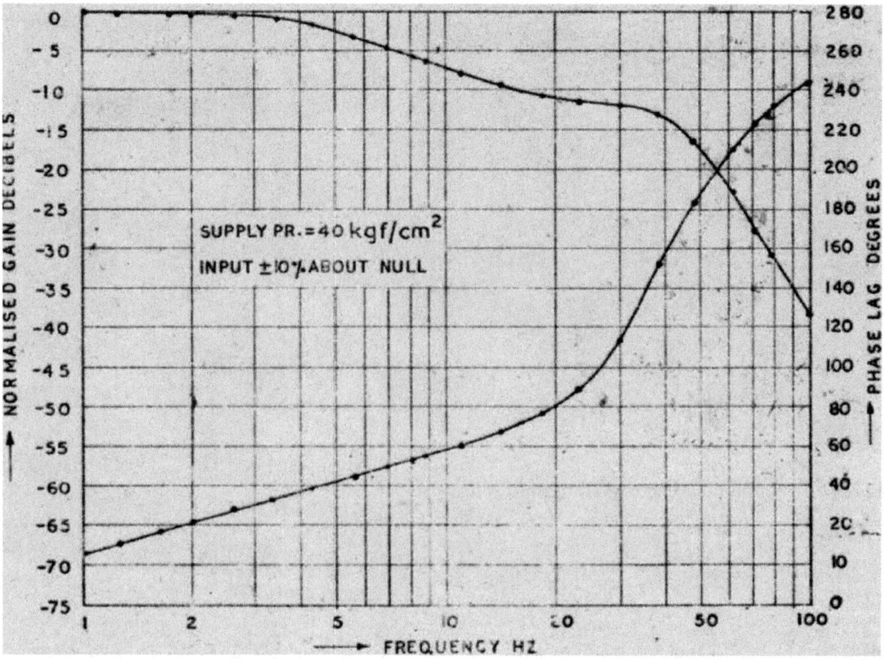

Fig. 4.22 Response measurement using frequency response analyzer

The response is also measured from a random noise signal Figs. 4.23, 4.24 and 4.25

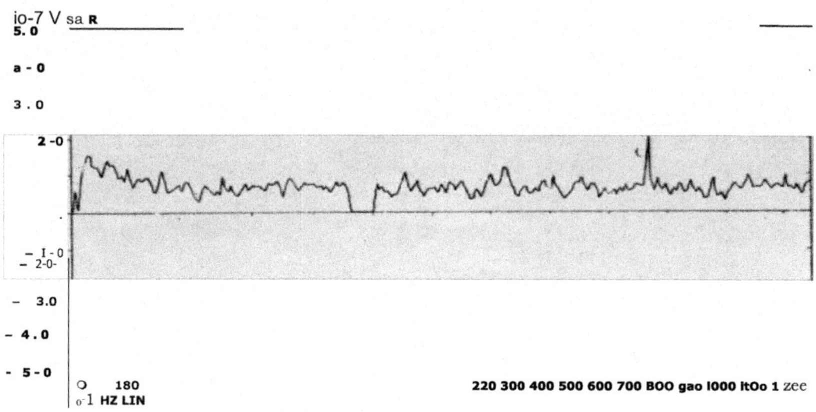

Fig. 4.23 Random input spectrum

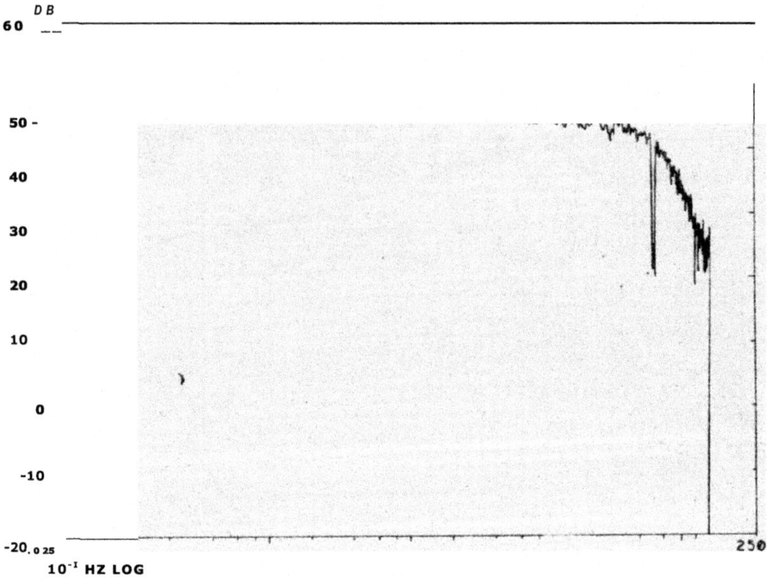

Fig. 4.24 Amplitude response from fourier analyzer

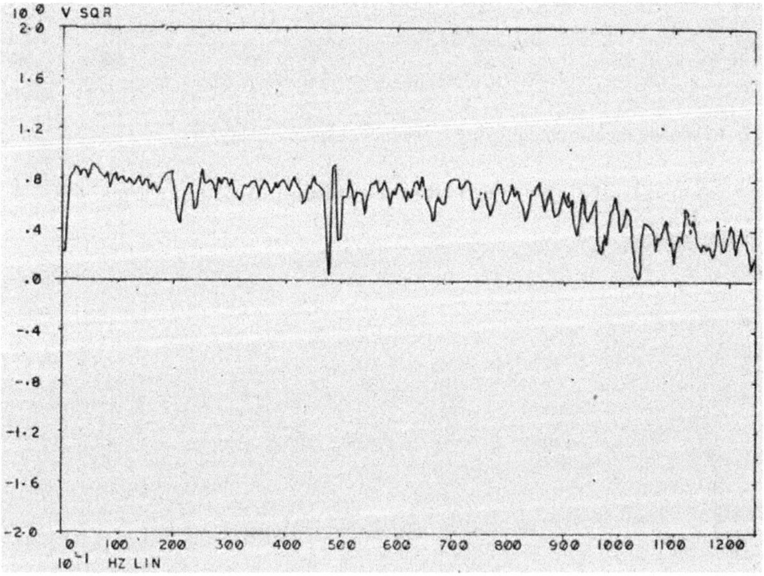

Fig. 4.25 Coherence function

The coherence function indicates the amount of output due to the input given. A value of 0.7 or more is taken to be acceptable.

Based on the measurements, the models are curves fitted from the analyzer. Fig. 4.26

Fig. 4.26 Bench testing servo drive packages with simulated inertia

Systems are also identified by subjecting them to the signal compression method (SCM). [6] presents the results of measuring the response of an Electrohydrostatic Actuator system in a closed loop position Figs. 4.27 and 4.28.

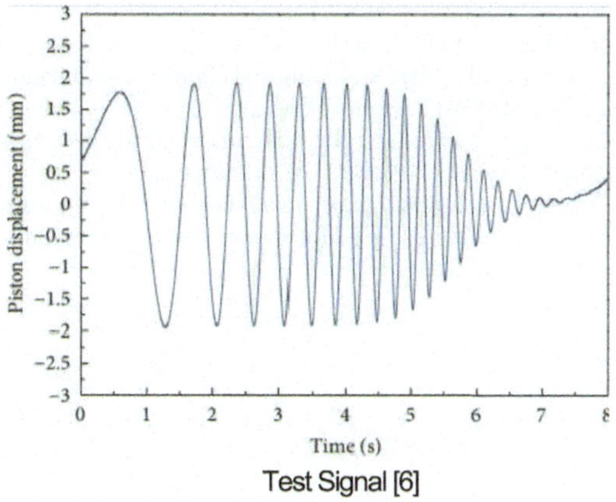

Fig. 4.27 Test signal [6]

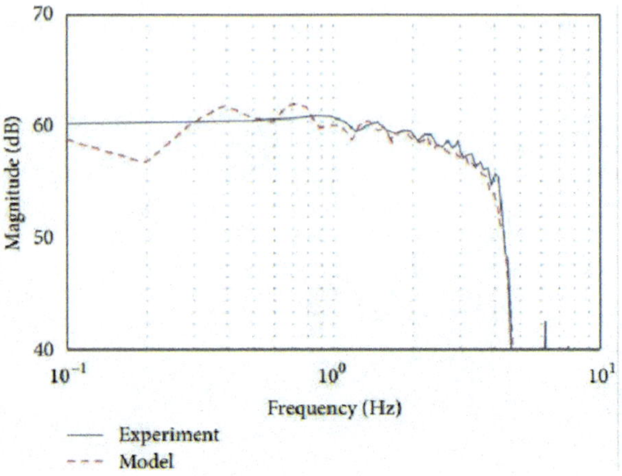

Fig. 4.28 Bode plots of the closed loop model and actual system [6]

References

1. Neal TP (1974) Performance estimation for electrohydraulic control systems. MOOG Tech Bull 126
2. Merritt HE (1967) Hydraulic control systems. Wiley
3. Kamil Afcan, Dr. Balkan T et al (2004) Modeling and control of stabilization systems for shaking tables. Master thesis. Middle East Technical University
4. Plummer A (2007) Control techniques for structural testing: a review. In: Proceedings of institution of mechanical engineers. Part1. J Syst Control Eng
5. Lee JM, Park SH, Kim JS Design and experimental evaluation of a robust position controller for an electro-hydrostatic actuator using adaptive anti windup sliding mode scheme. Sci World J 2013:16, article 590708, HINDAWI

Chapter 5
Control of SISO EH Servo Systems

Abstract Keeping in view the Control objectives and the challenges of non linearities, parameter variations and structured/unstructured Disturbances,a range of Control Options in Linear/Non linear regions are presented.These include the state variable feedback based LQG, Model based, Sliding Mode Variable Structure and the modern Intelligent Control methods.

5.1 General

The objectives of control design are as follows:

- Tracking of command signal accurately
- Robustness to variations in parameters
- Effective rejection of load disturbances, both structured and unstructured.

Different systems lay varying degrees of premium on each of these requirements. Cost functions vary. Challenges to achieve control objectives are the many system nonlinearities, parameter variations in the work envelope and load disturbances.

These objectives are met by a chain of open- and closed-loop systems. The feedback controller is designed after system identification to minimize errors due to nonlinear elements of the system ensuring stability under all operating conditions. The feedforward controller, which is ideally the inverse of the feedback controller, minimizes the dynamic errors and provides the desired robustness. The trajectory generator is known in most cases, while it is generated/derived in some cases (Fig. 5.1).

D. R. Raghavendra, *Electrohydraulic Servo Systems*, Springer Tracts in Mechanical Engineering, https://doi.org/10.1007/978-981-19-8065-7_5

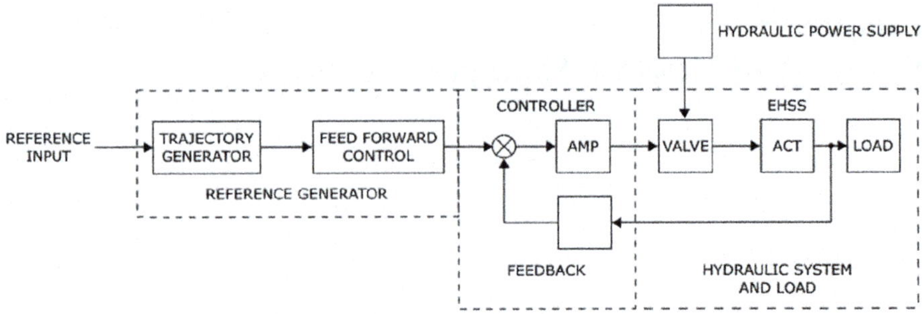

Fig. 5.1 Control elements of SISO systems

5.2 Challenges

System nonlinearities are the valve characteristics of lap conditions, flow/pressure characteristics, saturation, bulk modulus of oil and friction of the drive, including actuator and drive friction. The effects of these nonlinearities are individually studied by the describing function (DF) method in the frequency domain [1]. They tend to destabilize the system by shifting the critical Nyquist point or causing inaccuracies in control. Parameter variations within the work envelope and with use are common to most applications, while load disturbances are specific to some.

5.3 Control Design

The control design starts with a model of the system, a trial control option that is analyzed by simulation and testing for verification with specifications and iterating until the specifications are met (Fig. 5.2).

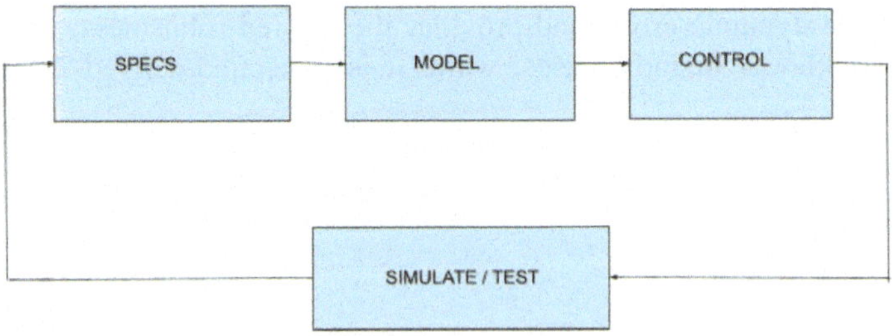

Fig. 5.2 Control design cycle

Three common approaches are used in modeling control systems:

- *White box Model*—based on the physical relationship of various parameters. Literature is abundant on linear and nonlinear models of servo valve actuators and other elements. The model can be linear, linearized, or nonlinear. There is no single model that represents the system in all details.
- *Gray box Model*—by identifying the parameters from tests. The servo valve coefficients, valve hysteresis, and prevailing friction conditions are measured to improve upon white box models.
- *Black box Model*—realized on input/output time histories, basically nonparametric.

System identification is performed by measurement under operating conditions online. Autoregression ARMA*x* models are generated by feeding the input, output and error functions. Suitable order models are curve-fitted to the actual model by linear/nonlinear least squares methods.

Powerful mathematical tools, Simulink, AMESIM, compatible with MATLAB enable conversion to other system representations: transfer function, pole zero or state space variable models (Fig. 5.3).

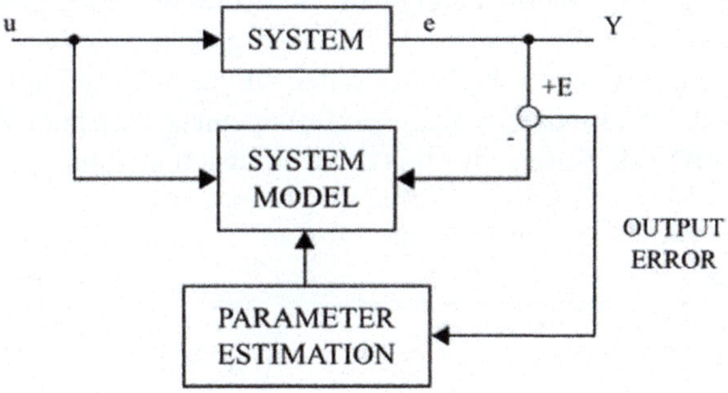

Fig. 5.3 System identification by minimizing output error (E)

Control options are many starting with PID control, which meets many specifications. Linearizing with additional inner loops to increase the accuracy/response by augmenting the damping factor is practiced in classical control theory-based systems. Linearizing with state variable feedback and modeling, applying powerful optimizing functions is done in digital/modern control systems. Linear, nonlinear LQG, adaptive control, model based or predictive model, self-tuning PID, variable structure, and H_2/H_∞ control tools are founded on modern control theory. Intelligent control methods such as fuzzy PID, neural networks, and iterative learning are under research [2].

An attempt is made in the following paras to cite examples of these control options applied to several EH servo systems.

5.4 Position Control

The classic frequency domain model of a servo valve-controlled balanced actuator driving a load is shown in Fig. 5.4.

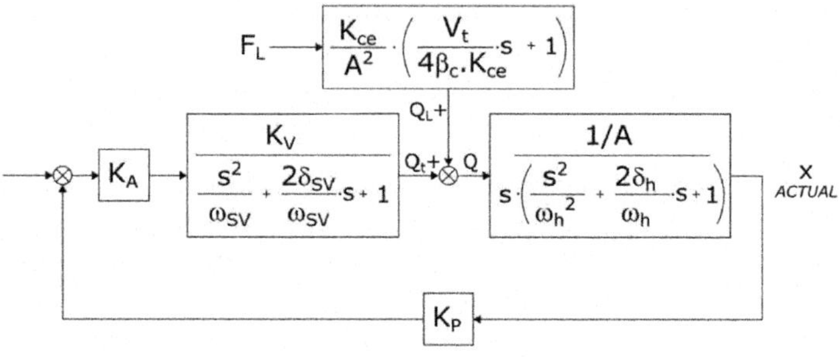

Fig. 5.4 Block diagram of a position control system

The inertial mass of load interacting with the hydraulic spring forming the poorly damped load/actuator resonance limits the loop gain K (product of $K_a \times K_v \times K_f/A$). Loop gain needs to be as high as possible within stability limits. Accuracy and closed-loop steady-state and transient response depend on loop gain. Ziegler–Nichols recommended PID gain settings for this drive are listed in Table 5.1.

Table 5.1 Ziegler–Nichols PID tuning recommendations

Type of controller	K_p	T_i	T_d
P	$0.5K_{cr}$	∞	0
PI	$0.45K_{cr}$	$\frac{1}{1.2}P_{cr}$	0
PID	$0.6K_{cr}$	$0.5P_{cr}$	$0.125P_{cr}$

K_{cr} is the threshold gain at which instability surfaces, and P_{cr} is the period of this oscillation. For P control, a gain of $K_{cr}/2$ ensures a minimum gain margin of 6 dB. The PID settings optimize the steady-state and transient response of the system and serve as a standard reference for all other control options.

Proper addition of velocity and acceleration feedback signals (state variables) locate zeroes close to the load actuator resonance poles. The additional damping enables higher Ku settings, which substantially increases the overall system performance.

The state variables velocity and acceleration can be derived without transducers using Luenberger observer models, which provide estimates of these parameters. These observers are now replaced by Kalman filters, which provide these estimates in the presence of noisy disturbance signals from load and displacement transducers.

5.5 Cascade Control

Cascade control relies on the output of primary position control to set a reference for the fast responding force control loop nestling the outer position loop. The controlled force drives the mechanical load to track the position command (Fig. 5.5).

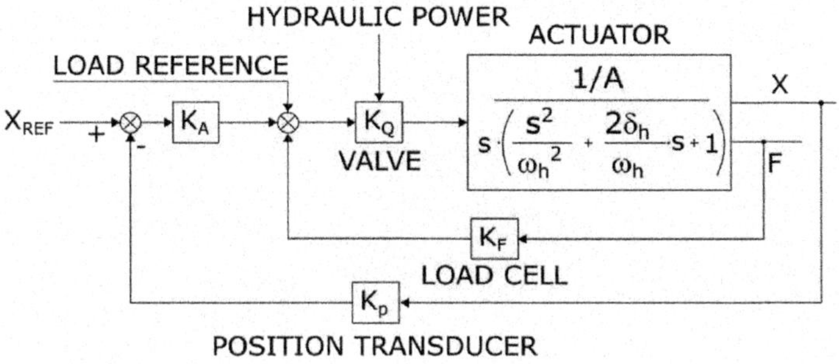

Fig. 5.5 Cascade control schematics

In applications where the exact moving mass is known, as in vibration testing, the force feedback can be used with feed forward acceleration control with good tracking performance.

5.6 Three Variable Motion Control

Vibration testing calls for position, velocity, and acceleration control referred to as three variable control (TVC) (Fig. 5.6).

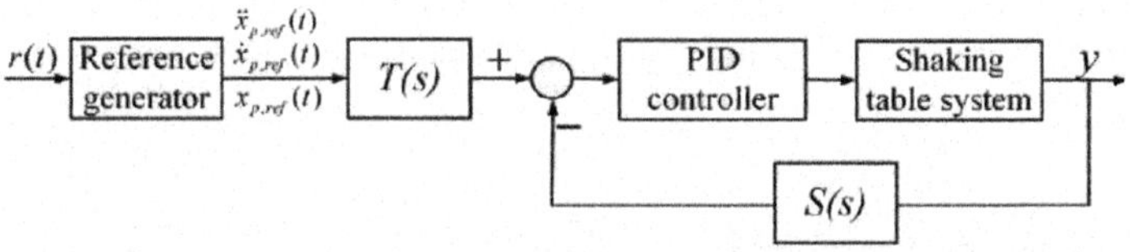

Fig. 5.6 Structure of TVC

Reference (Command) is generated from the known trajectory and the computed feedforward terms. Reference [3] notes that the velocity feedforward is scaled and added to the PI output. Additionally, in practice, position and acceleration signals are measured, and the velocity term for feedback is estimated using complementary filters.

Acceleration tracking suffers at high frequencies because of reduced position feedback and pronounced effects of measurement noise. Vibration tests call for constant acceleration over a frequency bandwidth scheme for acceleration tracking is presented in [4].

The acceleration command is generated offline in the feedforward section by passing the desired acceleration through a low-pass filter and the inverse of the approximated table position. The reference position command is then computed from a system dynamic compensator (Fig. 5.7).

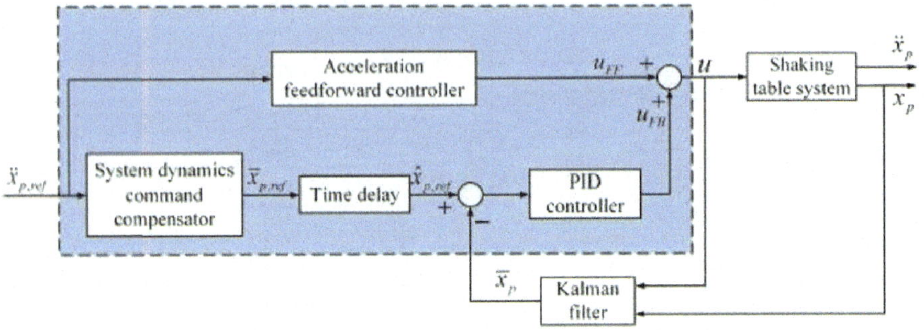

Fig. 5.7 Acceleration tracking, ATTC [4]

The control scheme uses the feed forward as the system driver and the feedback position loop as the stabilizer. Acceleration tracking is a challenge pursued by researchers armed with complex control schemes and adaptive LMS filtering [5].

5.7 LQG Control

LQG control is a state space technique to design a controller with integral action that allows a trade-off between tracking performance and control effort accounting for load disturbance and measurement noise.

The use of state variable feedback along with LQG optimization to mitigate oscillations during crane operation and to eliminate steady-state position error is described in [6] (Fig 5.8).

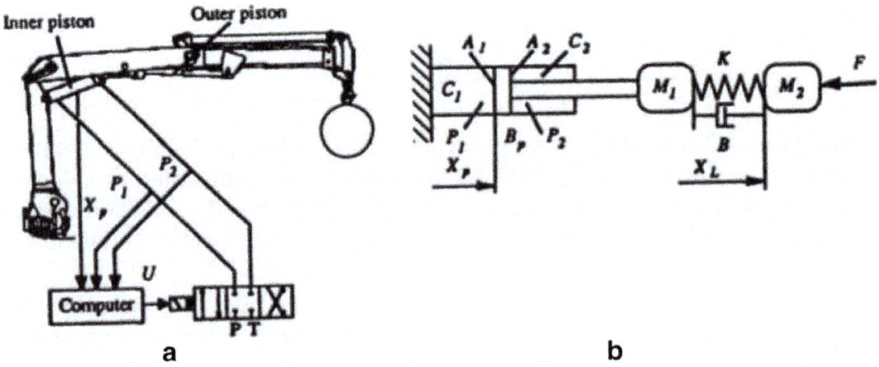

Fig. 5.8 **a** Hydraulic crane [6], **b** simplified circuit [6]

A model is developed considering the state variables—piston position x, piston velocity, load, servo valve pressure, and chamber pressure and the variable piston velocity is minimized by optimizing the gain k, an 8×2 matrix, by a Kalman filter. The control effectiveness is shown in Fig. 5.9 (Fig. 5.10).

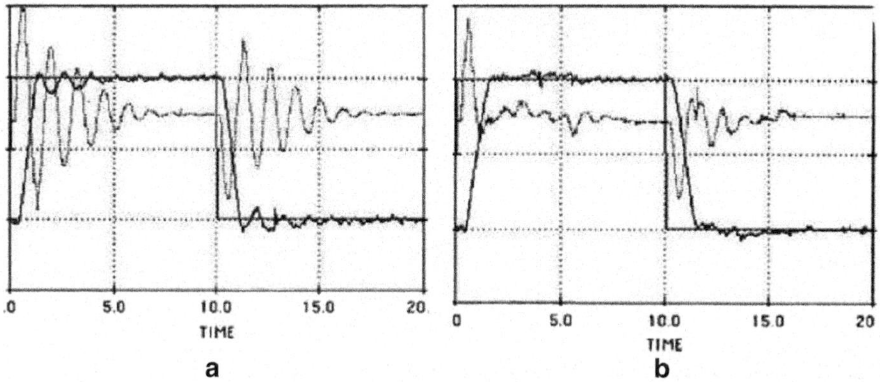

Fig. 5.9 **a** Response for load variation without LQG [6], **b** load response variation [6]

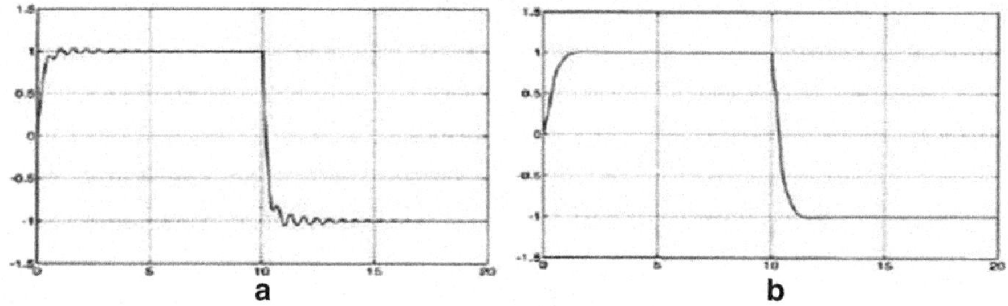

Fig. 5.10 Steady-state position without/with LQG [6]

5.8 Adaptive Control Methods

5.8.1 Model-Based Adaptive Control

A model, when fed with the system input and its output, continuously updates the controller parameters, thereby reducing errors.

On a black box model identified, two alternate approaches of adaptive control are compared for steering application [7]. The setup for the study is shown in Fig. 5.11.

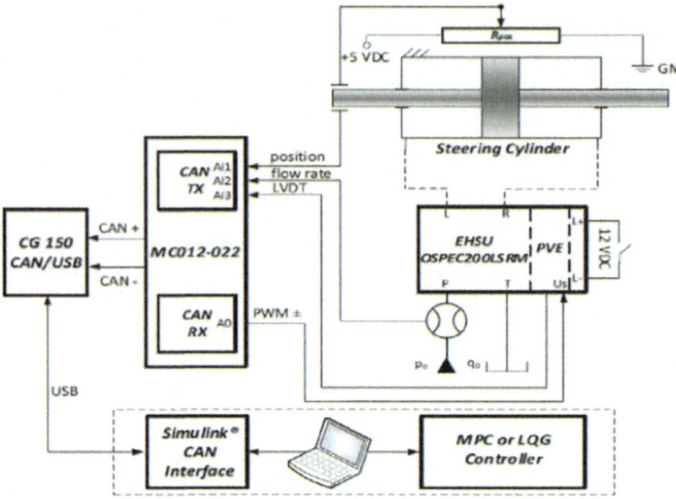

Fig. 5.11 Study of the layout of the steering system [7]

The system is identified by measuring the position of the piston, the spool position of the valve, and the flow for a random input to the system and the state variables identified by the Kalman filter for LQG control (Fig. 5.12).

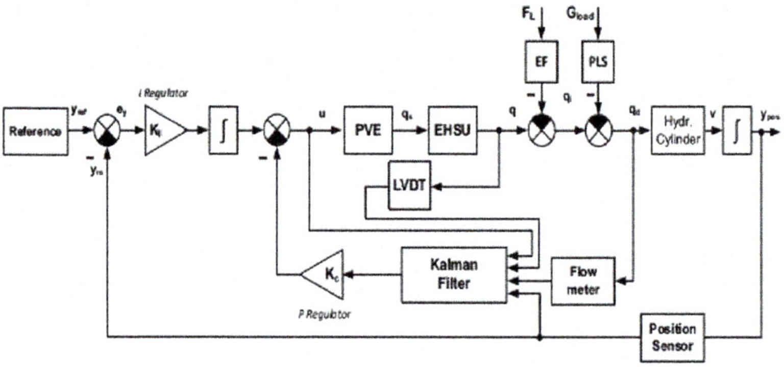

Fig. 5.12 Structure of LQG control [7]

The model predictive controller (MPC) is essentially a nonlinear model identified by a combination of step and swept frequencies. An autoregression model of order 6 is generated.

The relative performance of the two controllers is shown in Figs. 5.13 and 5.14.

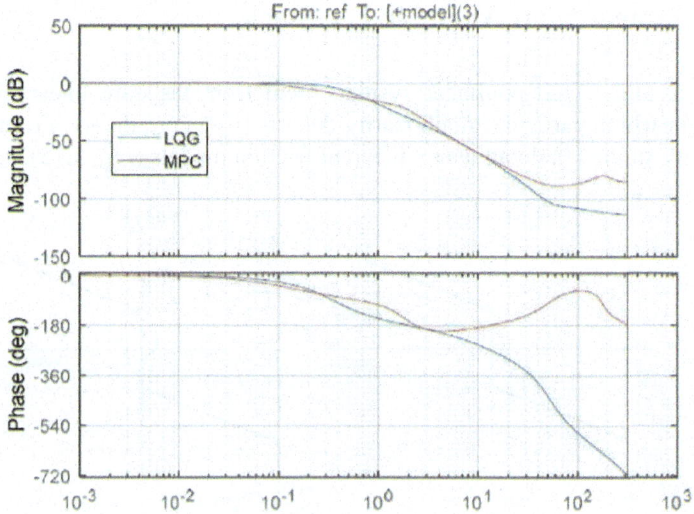

Fig. 5.13 Comparison of LQG with MPC [7]

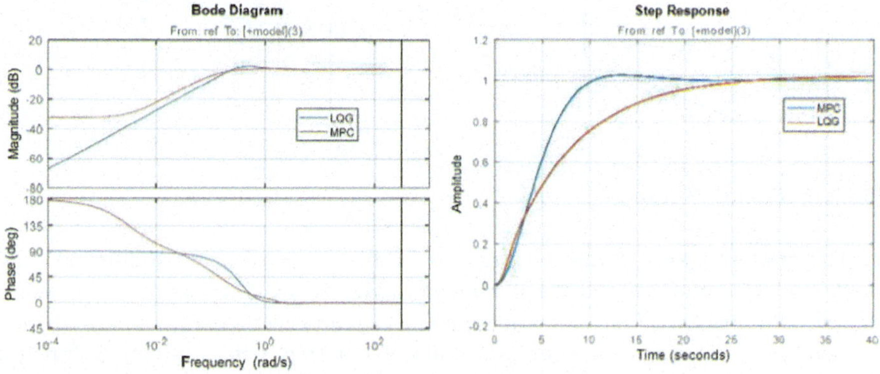

Fig. 5.14 Sensitivity to disturbance and step response comparison [7]

Authors note:

- Both controllers have comparable bandwidths, but the MPC controller has higher amplitudes at high frequencies, rendering them susceptible to measurement noise (Fig. 5.13).
- The LQG controller has better disturbance rejection (Fig. 5.14).
- LQG has better and faster response characteristics.
- LQG control has a better, faster response than the MPC controller.
- The LQG controller requires less energy to achieve the desired position.

5.9 Nonlinear State Variable Feedback

Methods to understand nonlinear systems stem from the core linear theory and the ways in which particular nonlinearity departs from core theory. Friction is one nonlinearity studied in detail using different friction models (Fig. 5.15).

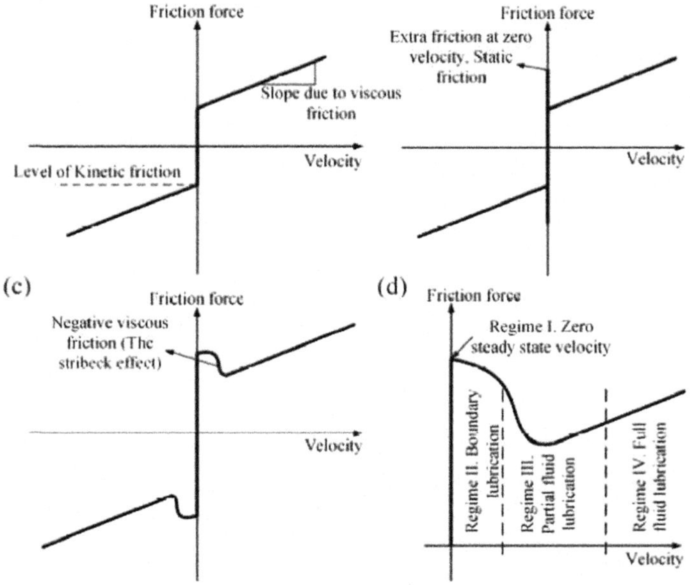

Fig. 5.15 Friction models

The discontinuous nature of friction is first approximated mathematically, rendering it continuous and the parameters assessed by measurements. They are linearized by backstepping and nonlinear least squares methods.

Reference [8] treats the simple Coulomb friction with a sigmoid function and details the results using a nonlinear backstepping approach. The results for exponential and sinusoidal inputs to a valve-controlled system are presented (Fig. 5.16).

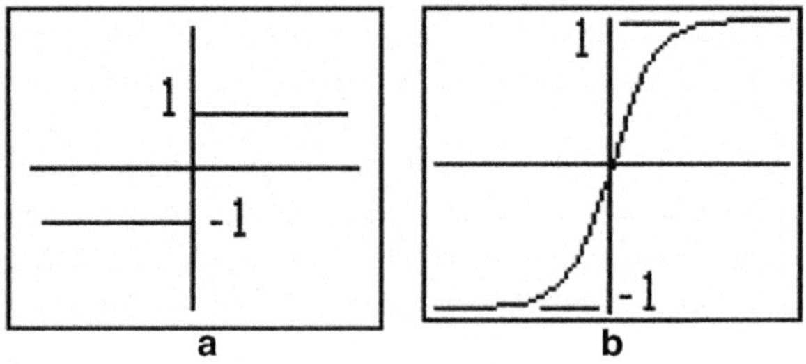

Fig. 5.16 **a** Friction model, **b** sigmoid function

Reduction of errors is apparent.

Reference [4] details the extraction of LuGre friction parameters by acceleration and position outputs for a low frequency sinusoidal input to a Servo valve controlled Vibration table.

5.9.1 Nonlinear Adaptive Robust Controller (NARC)

Reference [9] cites a NARC using backstepping technique and compares its performance with PID and "Determined Robust Controller" (DRC) (Fig. 5.17).

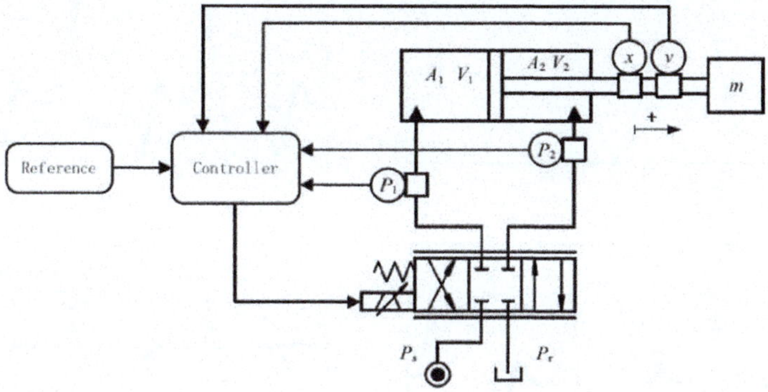

Fig. 5.17 Hydraulic system scheme for simulation study [9]

A set of state equations are derived, and an NARC is designed. Simulation studies of the performance of this controller are performed in the AMESIM and MATLAB platforms under parametric variations, nonlinearities and external disturbance conditions, and the results are presented (Figs. 5.18, 5.19, 5.20 and Table 5.2).

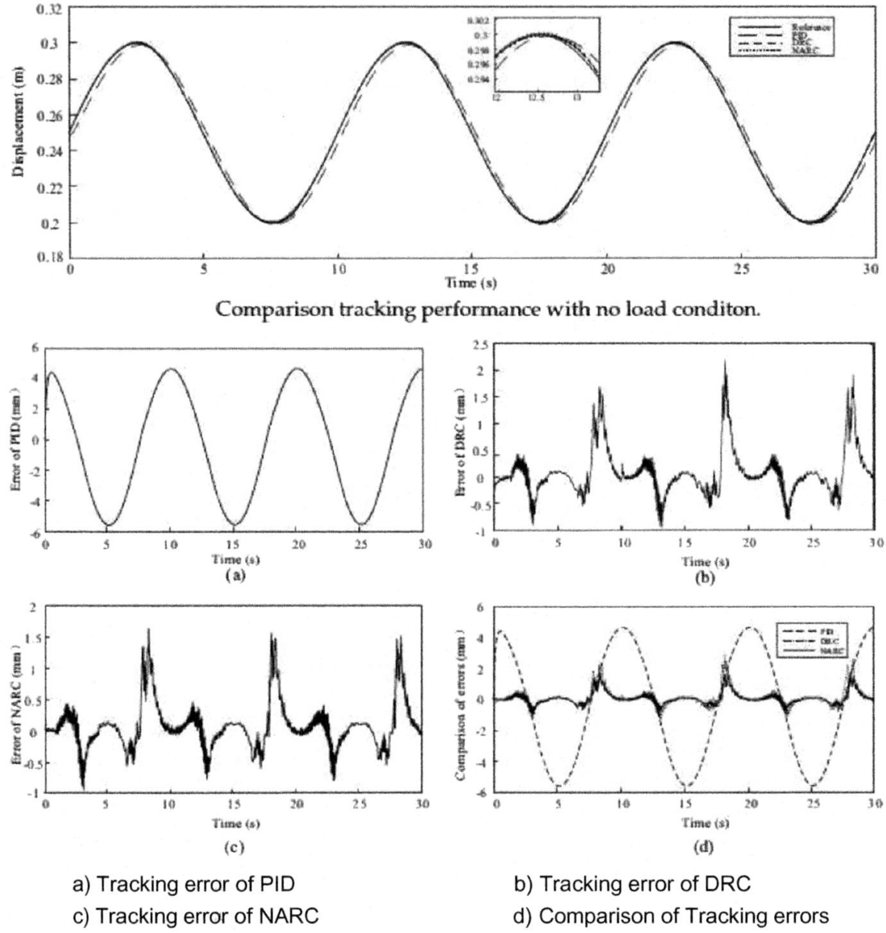

Fig. 5.18 Comparison under no load condition [9]

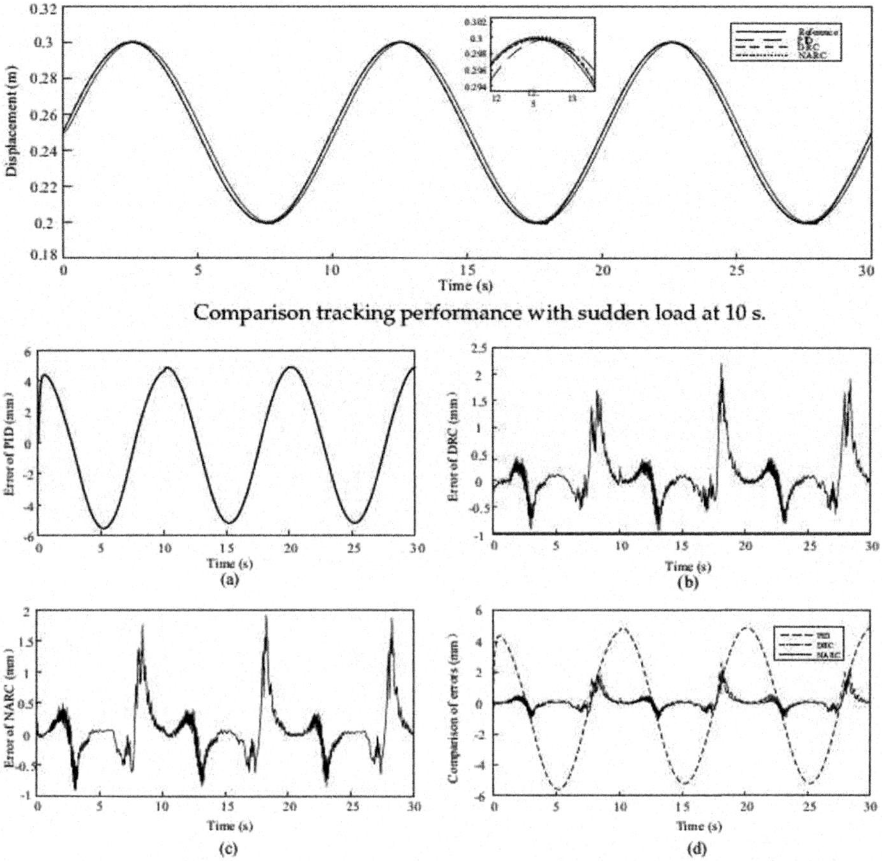

a) Tracking error of PID b) Tracking error of DRC
c) Tracking error of NARC d) Comparison of Tracking errors

Fig. 5.19 Comparison under sudden load conditions [9]

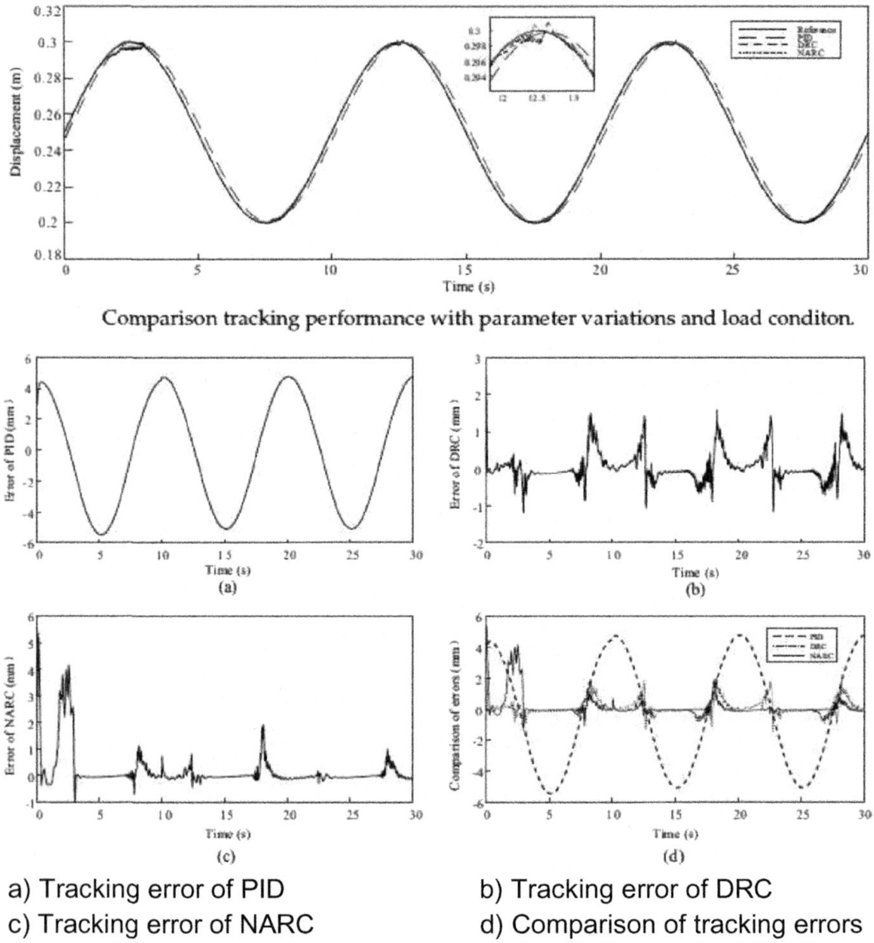

Fig. 5.20 Comparison with sudden load and parameter variation [9]

Table 5.2 Summary of NARC simulation [9]

Controller	Nominal condition			Sudden load condition			Variation and load condition		
	M/mm	μ/mm	ϕ/°	M/mm	μ/mm	ϕ/°	M/mm	μ/mm	ϕ/°
PID	4.7	3.2	5.4	4.9	3.2	5.9	4.8	3.2	6.1
DRC	2.7	0.27	0.07	2.4	0.275	0.0757	3.2	0.28	0.085
NARC	1.8	0.25	0.06	2.4	0.263	0.061	5.6	0.126	0.055

5.10 Wind Energy Harvesting

5.10.1 Pitch Servo System

Harvesting wind energy poses several control issues of maximizing power generation at wind speeds between threshold, 11 m/s and 25 m/s, providing safety at speeds beyond. The individual large radius Blade pitch control is subjected to large force fluctuations due to blade deflections. Reference [10] addresses this with novel control strategies for 1DOF and 4DOF blade models. The 4DOF blade model includes edgewise, flapwise, and torsional deflection of the blade with respect to the mean blade deflection, apart from the pitch angle [10] (Figs. 5.21 and 5.22).

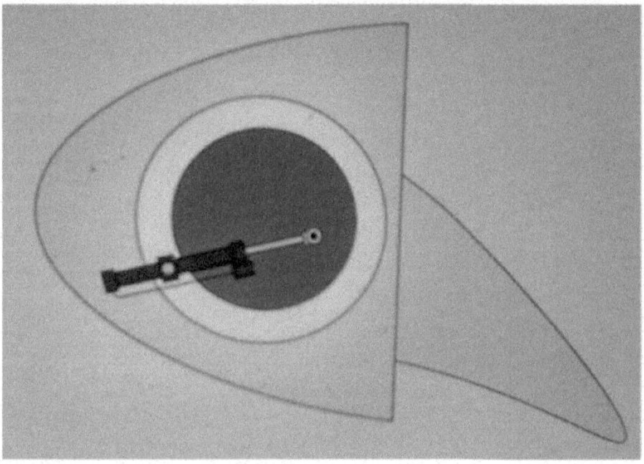

Fig. 5.21 Pitch servo system [10]

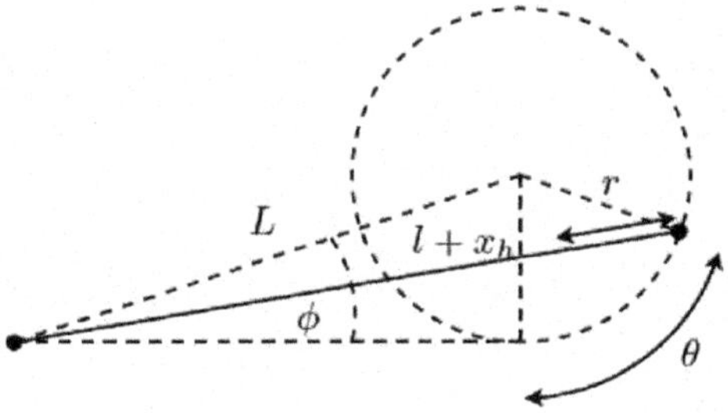

Fig. 5.22 Geometry of pitch actuator mechanics [10]

Energy-based Interconnection Damping Assignment and Passivity-based control laws are formulated along with a disturbance observer. The PD controller and a controller with a lead compensator where velocity measurements are not available are compared (Figs. 5.23, 5.24 and 5.25).

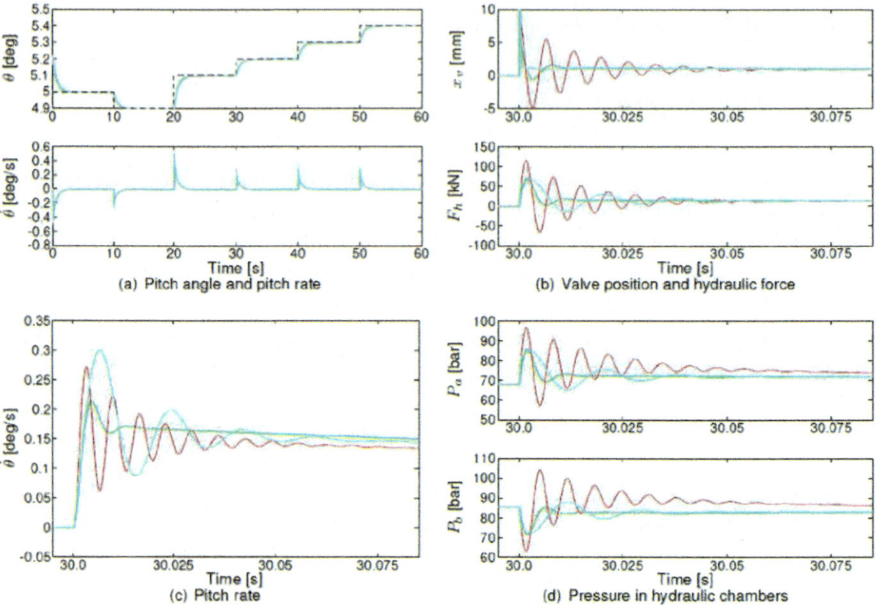

Fig. 5.23 Results for 1DOF [10]

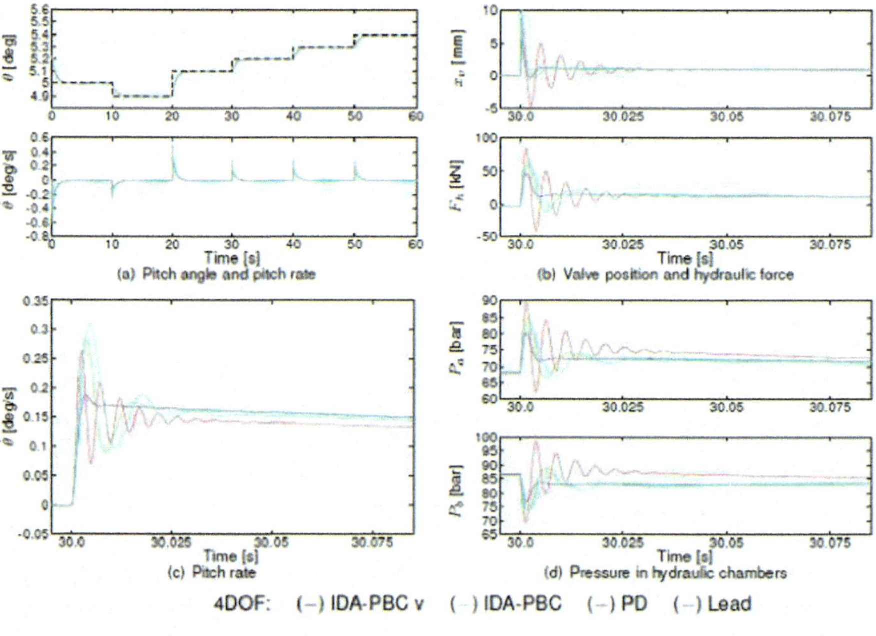

Fig. 5.24 Results for 4DOF [10]

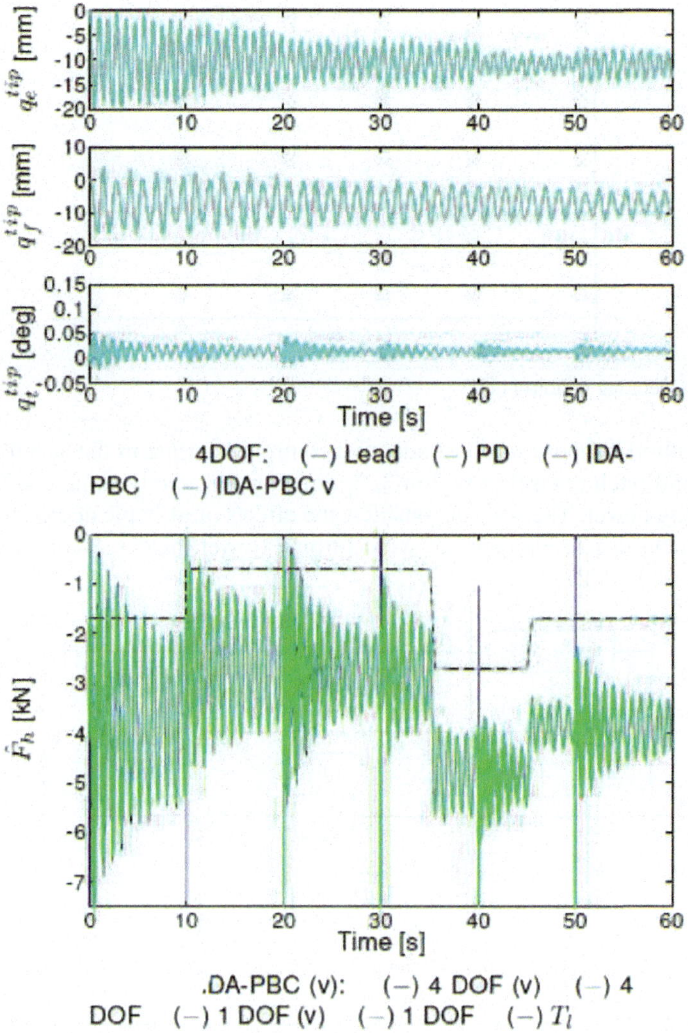

Fig. 5.25 Effect of load disturbances (fig). Extra 3DOF (top) [10]

5.10.2 Torque and Pitch Control

The objective of wind energy harvesting control is to maximize power generation under varying speeds in regions of 1 and 2 and provide safety in region 3 under stall conditions are addressed in [11] (Fig. 5.26).

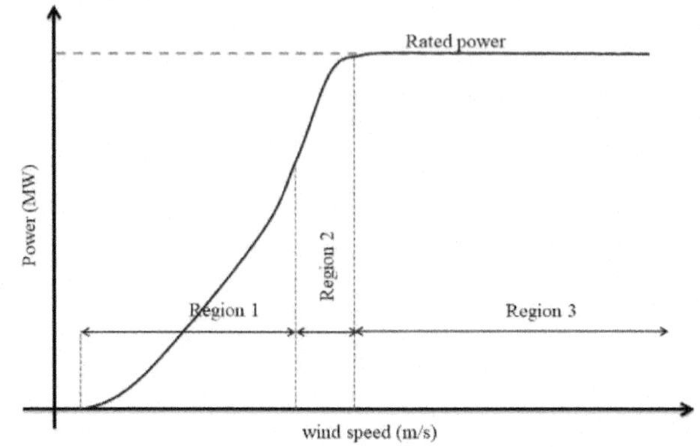

Fig. 5.26 Operating regions [11]

The authors propose a robust adaptive control of torque in the speed regions of 1 and 2 and pitch control in region 3, optimizing the aerodynamic coefficients for maximizing power. The control considers the effects of multiple degrees of freedom of movement due to elements of wind turbine (tower, blades, rotor, gearbox, and generator) (Fig. 5.27).

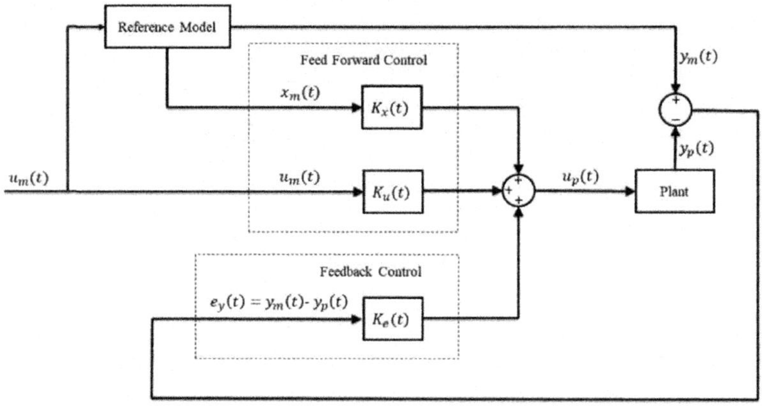

Fig. 5.27 Control schematics [11]

The adaptive control model is designed and implemented on the 2 MW and 5 MW wind turbine models (Fig. 5.28).

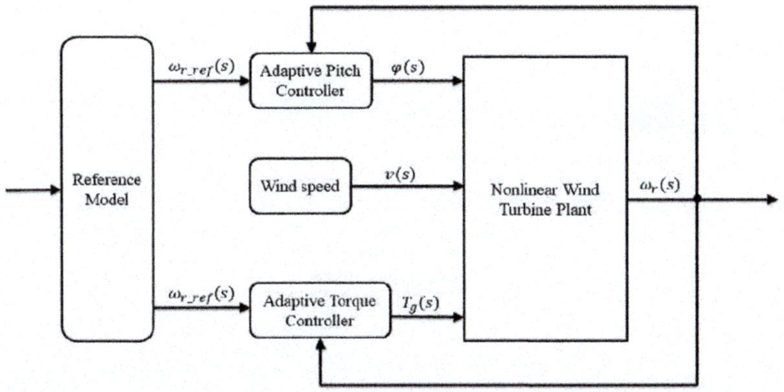

Fig. 5.28 Adaptive control module [11]

The performance of the adaptive control scheme on the two wind turbine models is analyzed using MATLAB/Simulink for three cases of pitch control at 16 m/s and torque control at 8 m/s with a standard deviation of 2 and for combined control over the speed range of the generator speed. The efficacy of control for all cases is shown in Fig. 5.29 (Figs. 5.30, 5.31, 5.32, 5.33 and 5.34).

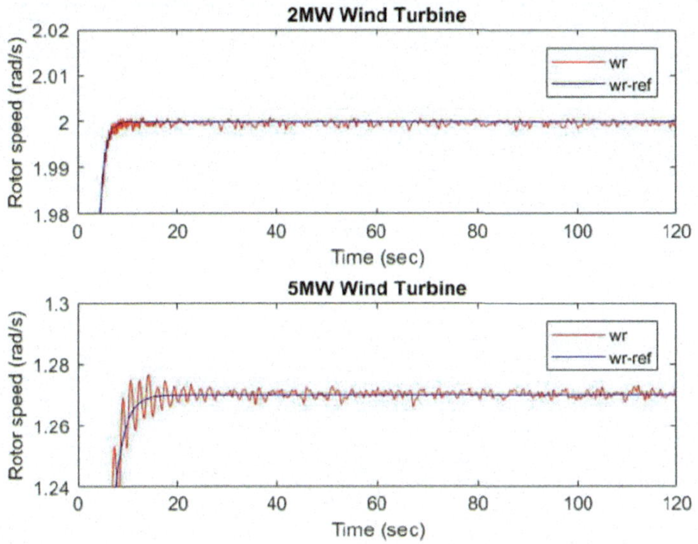

Fig. 5.29 Control efficacy [11]

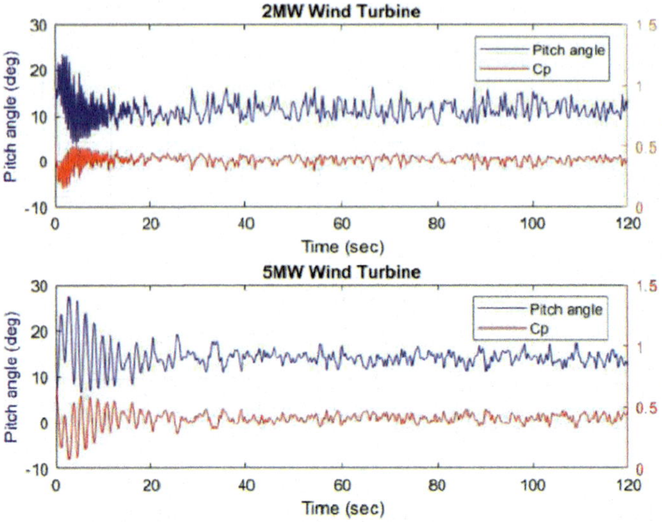

Fig. 5.30 Pitch angle control [11]

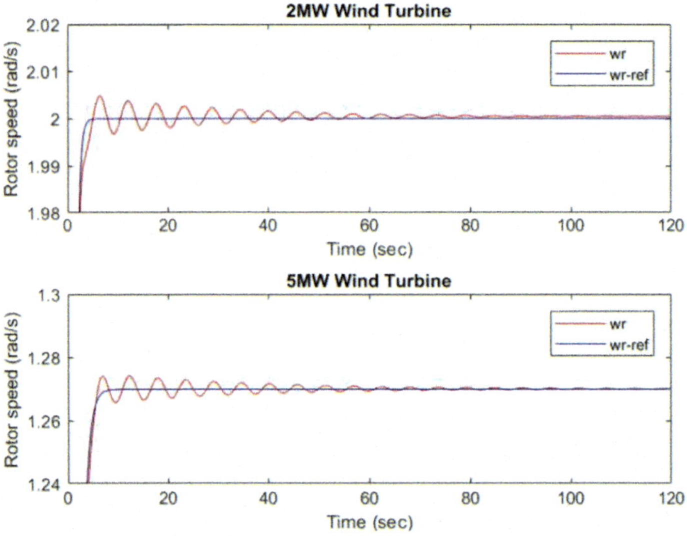

Fig. 5.31 Rotor speed control [11]

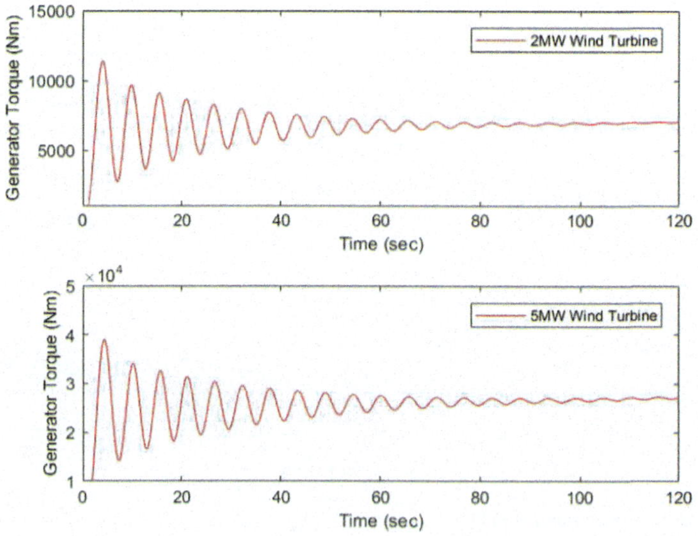

Fig. 5.32 Torque control [11]

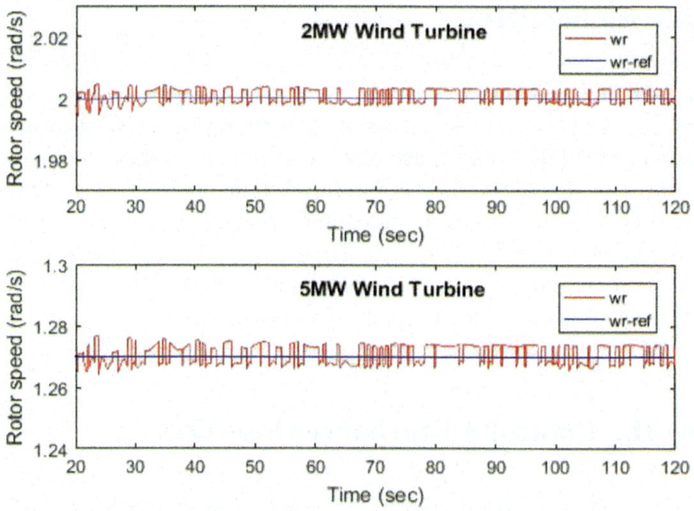

Fig. 5.33 Rotor speed control [11]

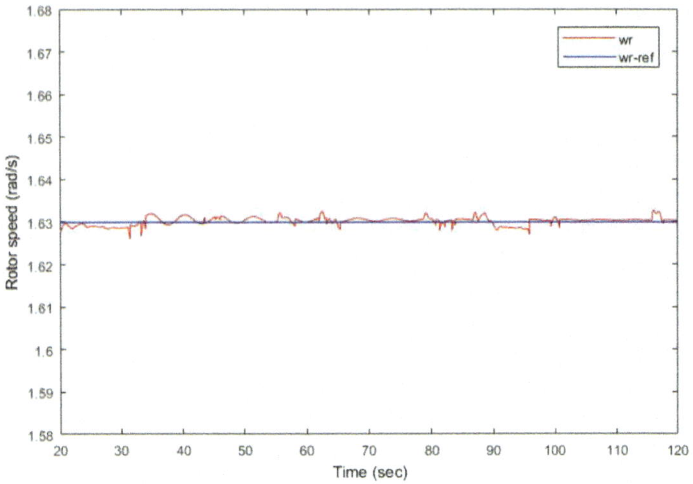

Fig. 5.34 Rotor speed control [11]

5.11 Self Tuning PID

The Ziegler–Nichols method of setting optimal PID settings is based on the measurement of the gain factor (K_{cr}) and period of oscillation (P_{cr}) at the threshold of instability. Reference [12] describes a discrete measurement of these parameters, which forms the basis for self-tuning PID control. Reference [13] suggests an incremental self-tuning controller based on a nonparametric black box model of the system. The ARMAX model obtained from the recursive least squares algorithm generated from past values of inputs and outputs. The paper details a method of solving the unknown controller polynomials from the measured system model.

5.12 H_2/H_∞ Control, Disturbance Observers

H_2/H_∞ Control: H_2/H_∞ norms address the conflicting requirements of disturbance rejection and sensitivity to parameter variations to which systems are subject. The H_2 norm aims to minimize the cost arising out of control input signal and Bode amplification or magnitude. The H_∞ norm fixes the magnitude of the damping factor over all frequencies. The mixed H_2/H_∞ norms optimize stability and rejection of disturbances [14]. System load disturbance (process noise) generally occurs at a low frequency, while measurement disturbance (measurement noise) occurs at a high frequency. Noting these contrasting design requirements, the H_2/H_∞ infinity controller designs an input prefilter considering the range of input operating frequencies. This input filter is in essence an inverse of the desired sensitivity and has an integrator.

5.12.1 LQG, H_∞ Comparison

Reference [15] presented an approach for robust stability and robust performance for the control of electrohydraulic power steering control for LQG and H_∞ control strategies (Figs. 5.35 and 5.36).

Fig. 5.35 Electro hydraulic power steering test rig [15]

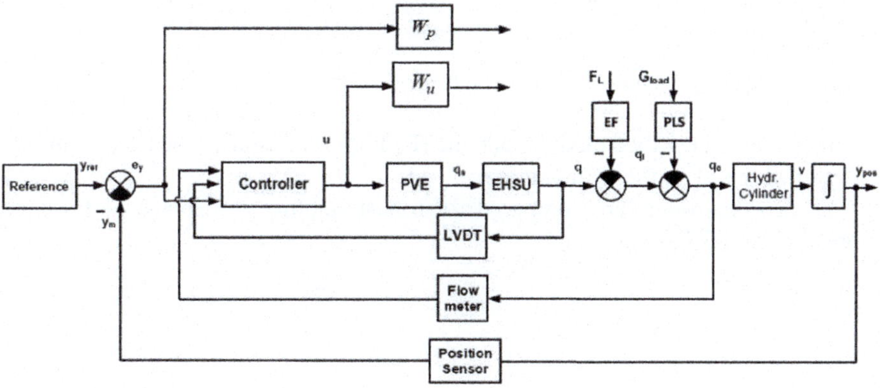

Fig. 5.36 Structure of control system [15]

For the Control input, Frequency responses of output valve spool position (y spool), measured flow rate (flow), and cylinder position (y, pos) are measured, and a black box plant model is derived (Figs. 5.37 and 5.38).

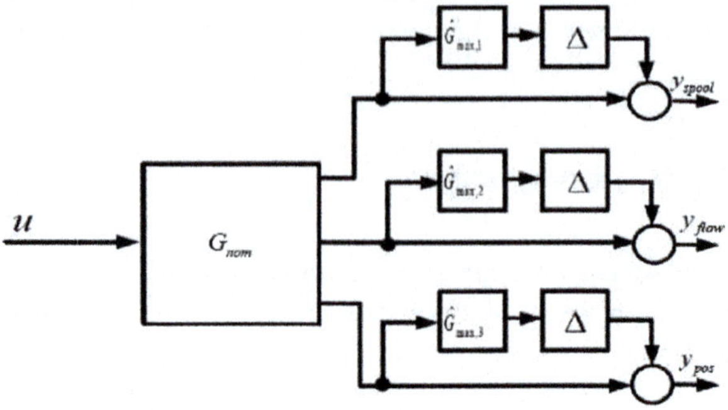

Fig. 5.37 Model with output multiplicative uncertainty [15]

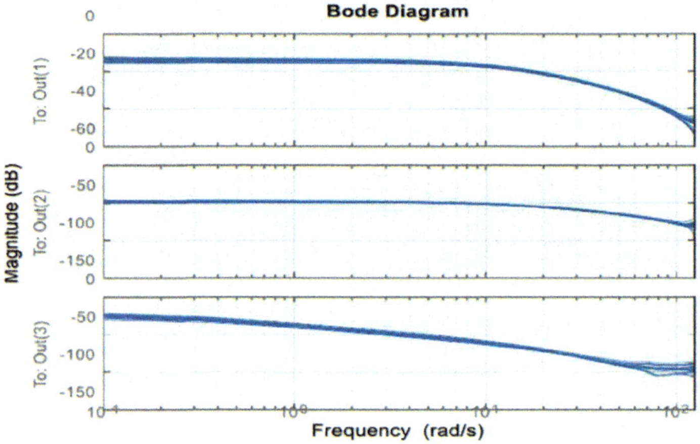

Fig. 5.38 Frequency responses with uncertainty mode [15]

Based on the identified plant model, the H_{∞} filter is designed. The LQG controller is designed based on the Kalman filter.

The two controllers are compared for robust stability and robust performance (Figs. 5.39 and 5.40).

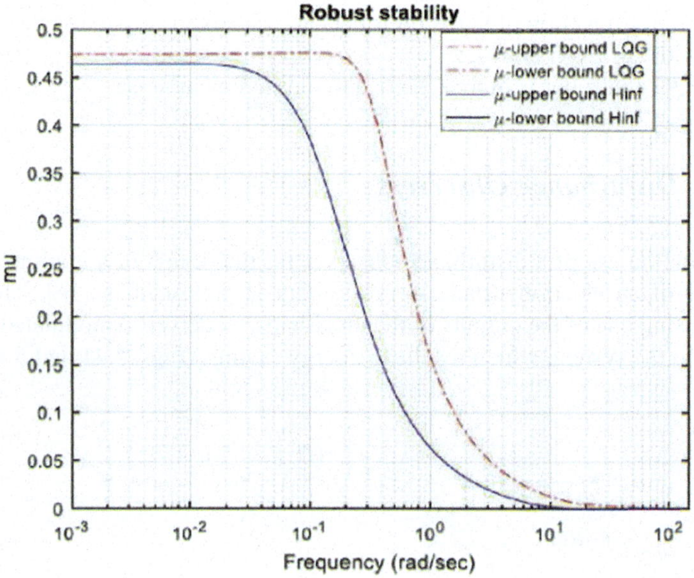

Fig. 5.39 Robust stability [15]

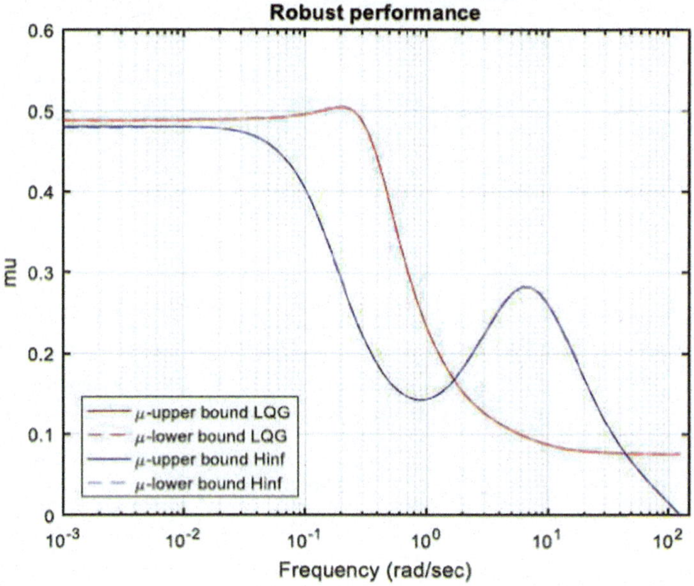

Fig. 5.40 Robust performance [15]

The authors present the output of two controllers for fifty random values of plant uncertainty.

They conclude:

- Disturbance rejection of both controllers are approximately the same.
- Bandwidth of LQG is wider.
- The LQG controller is more sensitive to measurement noise.

5.12.2 Disturbance Observers

A study on the use of a disturbance observer for the design of a nonlinear cascade controller of an electrohydraulic servo system is presented in [16]. The authors propose using one of the outputs of the disturbance observer to modulate the supply pressure to the system, which results in energy savings (Figs. 5.41 and 5.42).

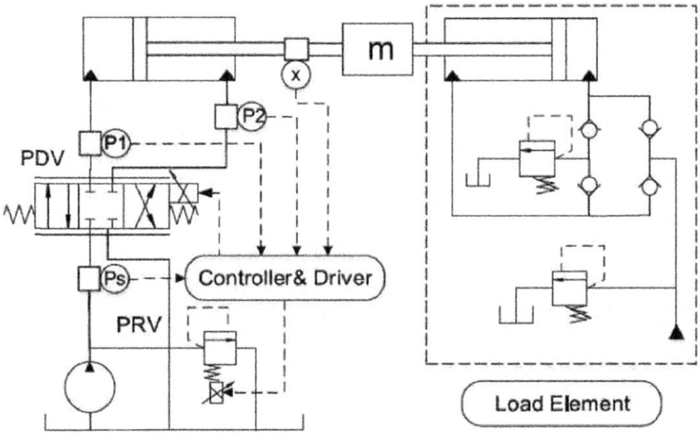

Fig. 5.41 System under study [16]

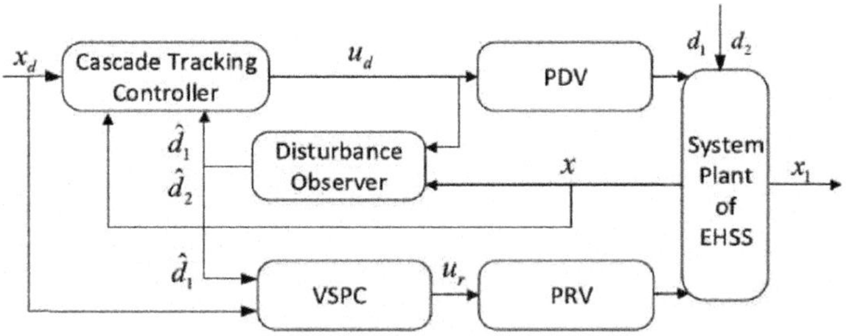

Fig. 5.42 Control structure with disturbance observer [16]

The design of the disturbance observer ensures that estimates of disturbances and their derivatives are bounded. Nonlinear cascade control combines PI control with a recursive backstepping procedure.

The experimental verification of the controller for the system was done with position transducer and pressure transducers for chamber pressures and system pressure (Figs. 5.43, 5.44, 5.45, 5.46 and Table 5.3).

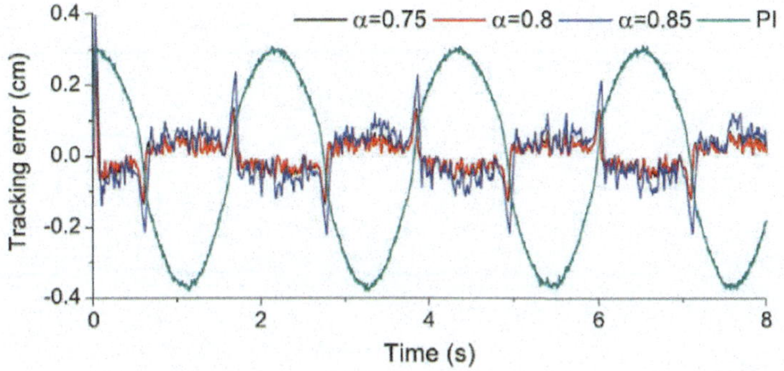

Fig. 5.43 Performance at different PI settings [16]

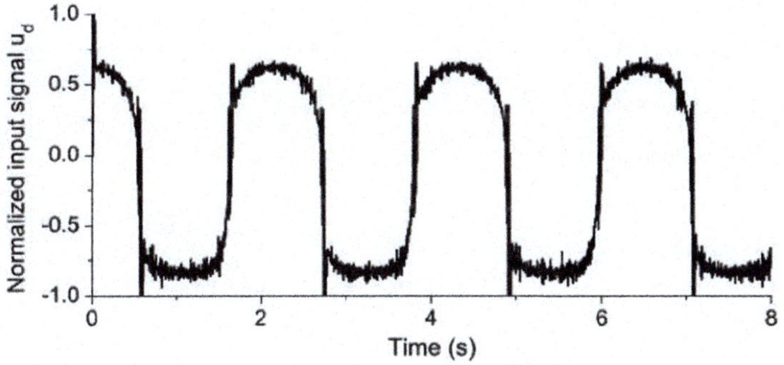

Fig. 5.44 Control input to valve [16]

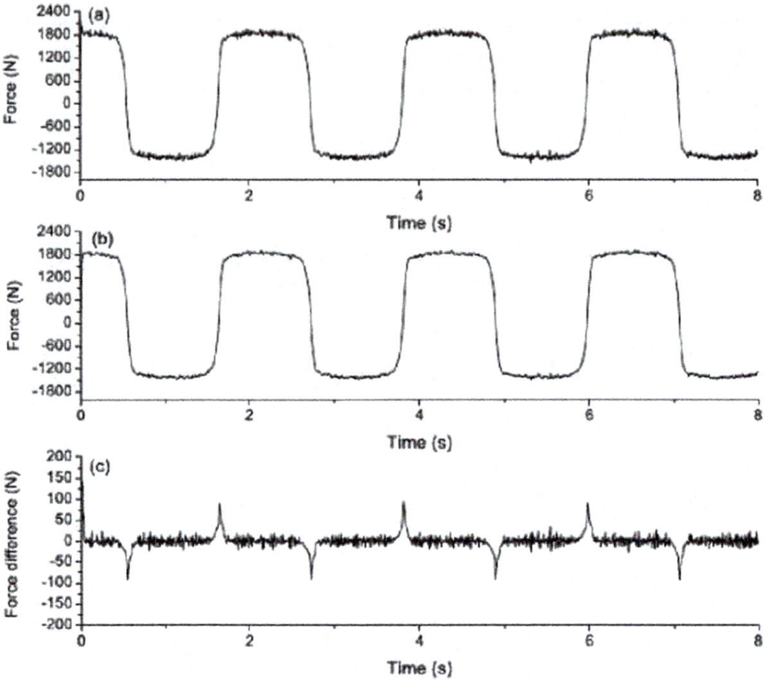

Fig. 5.45 Force estimation [16]

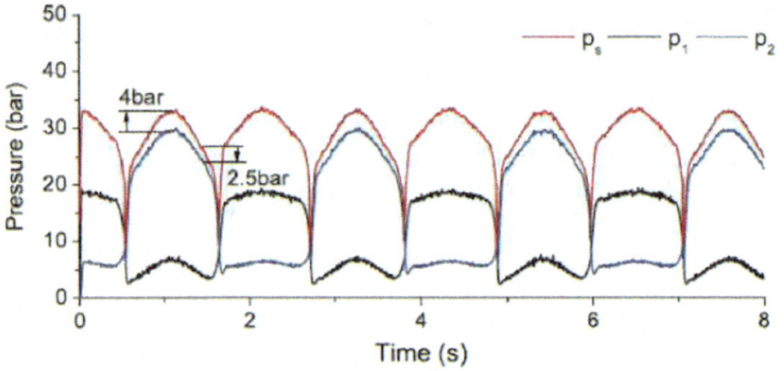

Fig. 5.46 Pump and chamber pressures during the harmonic test [16]

Table 5.3 System parameters [16]

Total leakage coefficient	3×10^{-11}	Oil bulk modulus	1×10^4 bar
Equivalent mass of the load	9.5 kg	Environment stiffness	0
Piston/rod diameter	40/30 mm	Cylinder stroke	20 cm
Oil density	890 kg/m^3	Viscous damping	800 N/m
Δp_{loss}	1.2 bar	k_{rv}	21.3 bar/V
$k_q k_x$	8.8×10^{-7}	λ	400
ε_1	8.2×10^{-5}	ε_2	1×10^4
k_1	500	k_2	800

5.12.3 *Use of Two Disturbance Observers*

Reference [17] recommends using two disturbance observers: one based on chamber pressure measurements for load and the other based on output position using an accurate differentiator (Figs. 5.47, 5.48 and 5.49).

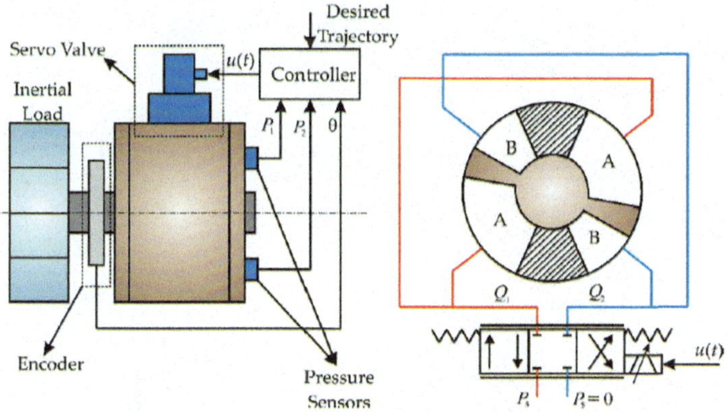

Fig. 5.47 System under study [17]

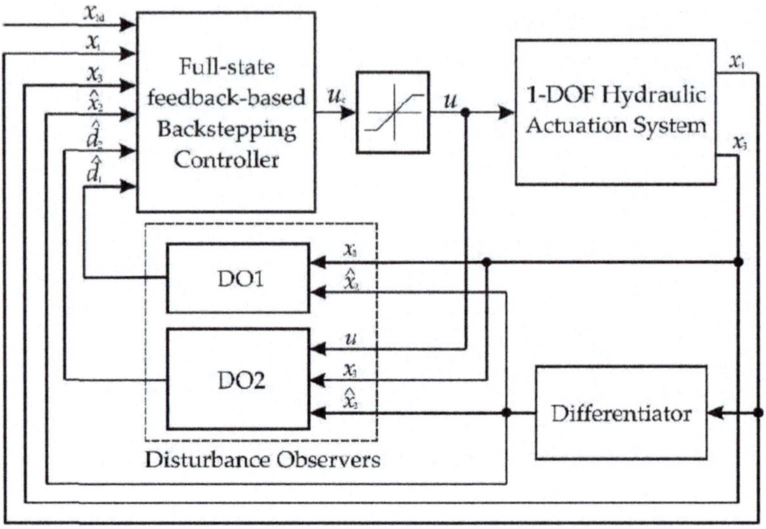

Fig. 5.48 Schematics of the proposed control system based on disturbance observers with backstepping control [17]

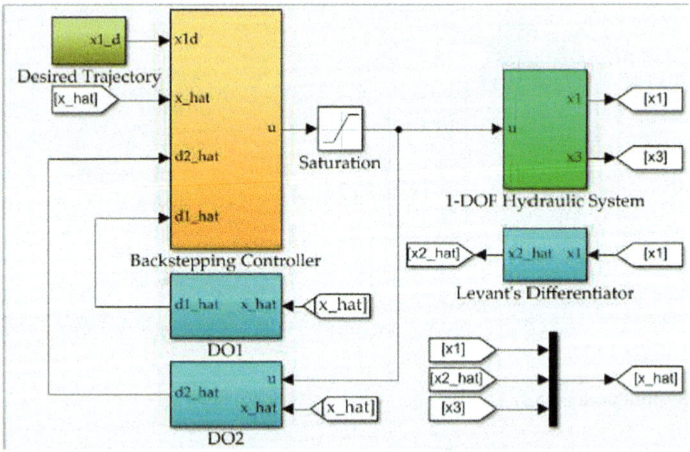

Fig. 5.49 Control in the MATLAB/simulink environment [17]

The control was compared with an *Extended State Observer with Backstepping Control* (ESOBC), two extended state observers with Backstepping Control (2ESOBC) and *Unknown Disturbance Estimation based Backstepping Control* (UDEBC).

The tracking performance of all controllers is shown in Fig. 5.50 (Fig. 5.51 and Tables 5.4, 5.5).

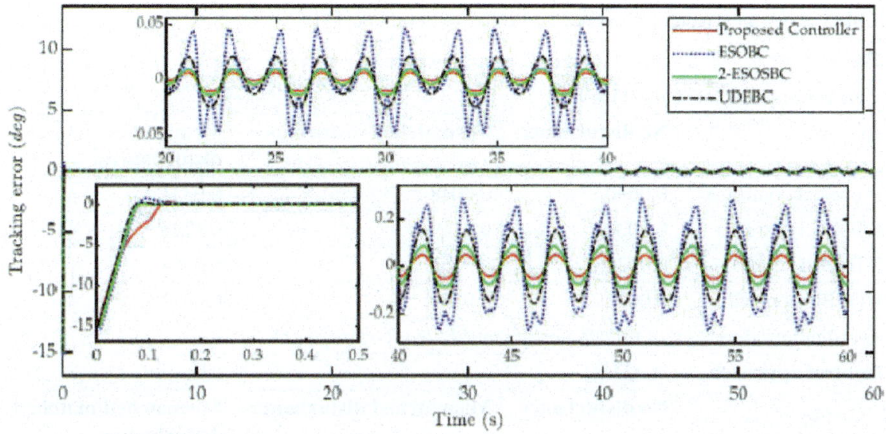

Fig. 5.50 Tracking errors of the proposed controllers and others in the presence of matched and mismatched disturbances [17]

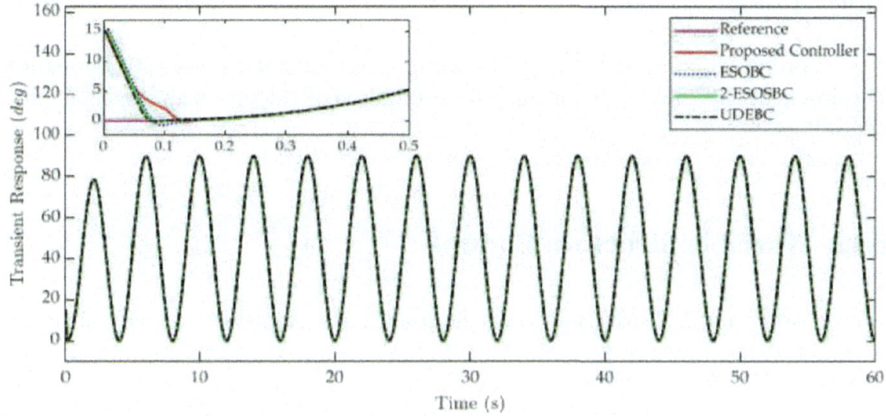

Fig. 5.51 Tracking performance of the proposed controller compared with others [17]

Table 5.4 Maximal absolute values of tracking errors in steady state [17]

Control approach	M_e (Deg)		
	No disturbance	Mismatched disturbance	Mismatched/matched disturbances
Proposed controller	0.0045	0.0093	0.0521
ESOBC controller	0.0045	0.0500	0.2860
2-ESOBC controller	0.0045	0.0123	0.0895
UDEBC controller	0.0057	0.0236	0.1572

Table 5.5 Comparison of proposed controller with others [17]

The average tracking errors in the steady-state phase			
Control approach	μ_e **(Deg)**		
	No disturbance	**Mismatched disturbance**	**Mismatched/matched disturbances**
Proposed controller	0.0030	0.0048	0.0331
ESOBC controller	0.0030	0.0219	0.1728
2-ESOBC controller	0.0030	0.0065	0.0584
UDEBC controller	0.0038	0.0131	0.0997
The standard deviation of tracking errors in the steady-state phase			
Control approach	σ_e **(Deg)**		
	No disturbance	**Mismatched disturbance**	**Mismatched/matched disturbances**
Proposed controller	0.0014	0.0030	0.0151
ESOBC controller	0.0014	0.0189	0.1698
2-ESOBC controller	0.0014	0.0038	0.0254
UDEBC controller	0.0018	0.0068	0.0469

The proposed controller deploys a Levant differentiator for velocity estimation. The article details the improvement in the estimation of velocity with respect to other controllers.

5.13 Variable Structure Control

Sliding mode control (SMC) addresses the control of systems with relay-type discontinuous nonlinearities. Sliding mode control uses a discontinuous input to force the system trajectory to a well-defined sliding surface (Fig. 5.52).

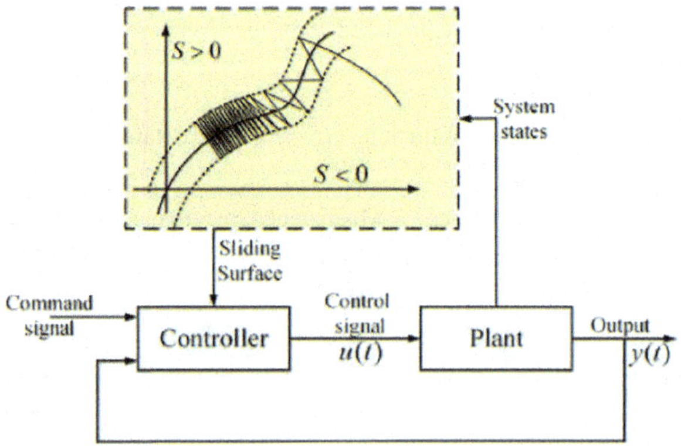

Fig. 5.52 Schematic of SMC control

SMC design is performed in two stages: switching function to develop the sliding surface and design control laws to force the system to follow the desired sliding surface [2].

5.13.1 SMC Application 1

Reference [18] details the application of SMC design and its verification of robustness for different masses (Fig. 5.53).

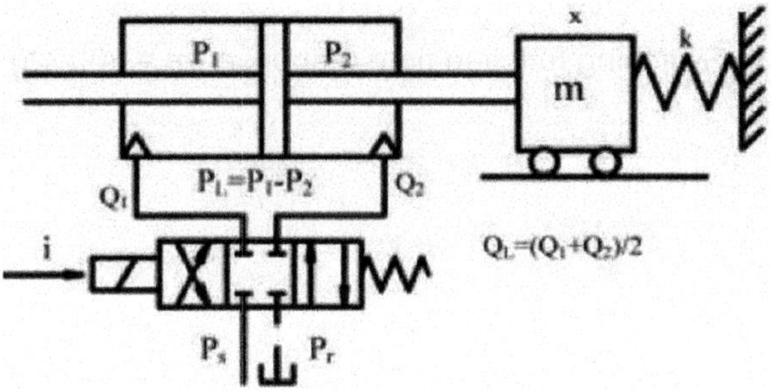

Fig. 5.53 Hydraulic scheme

The design uses an inverse tan function for switching to reduce oscillations due to discontinuous inputs (Fig. 5.54).

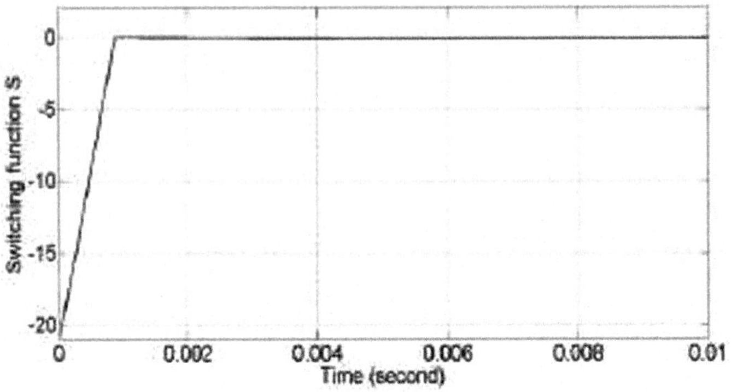

a) Switching function time response (m = 200 kg)

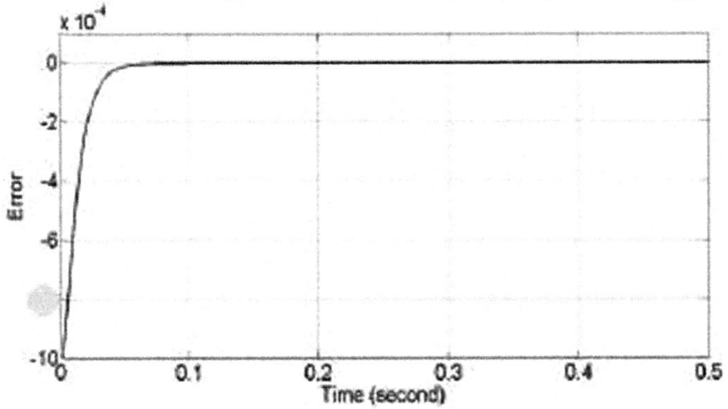

b) Error response$x_1(t)$

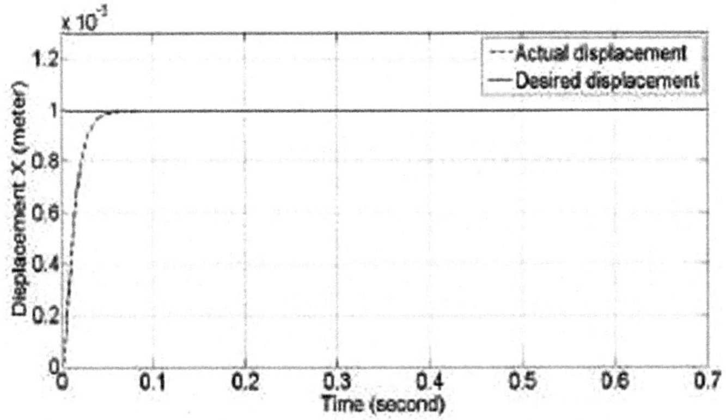

c) Position of system $x(t)$ (responseto unit step input) (m = 200 kg)

Fig. 5.54 Asymptotic response to switching function [18]

The robustness of control is demonstrated with a different mass (Figs. 5.55 and 5.56).

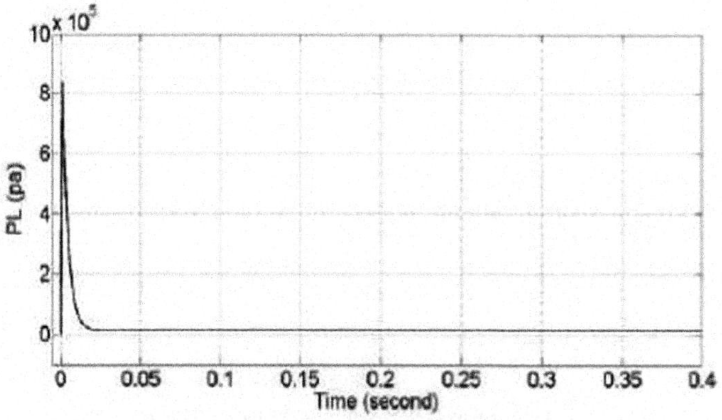

(a) Load pressure P_L time response

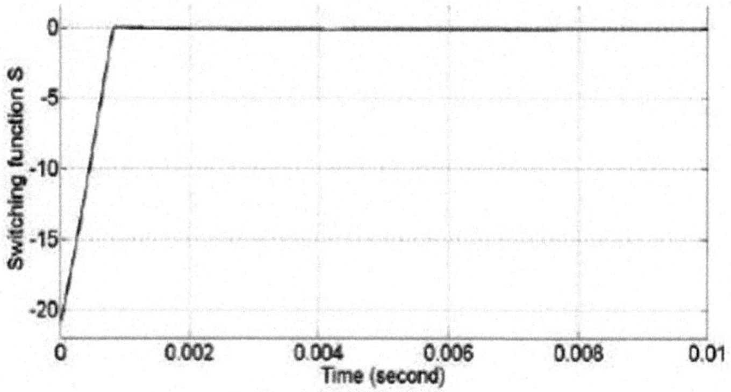

(b) Switching function time response (m=75kg)

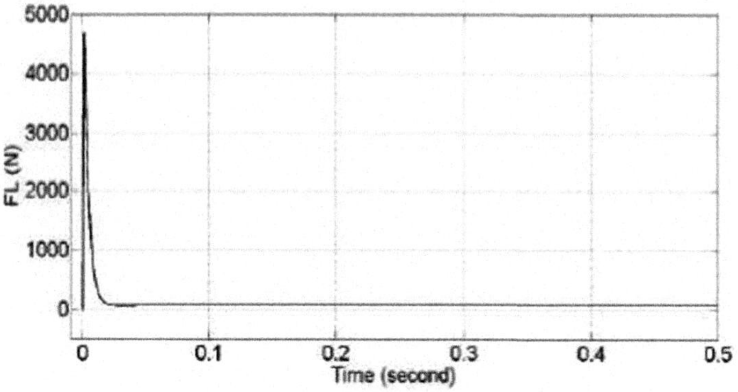

(c) Hydraulic force time response (FL = PL * Ω)

Fig. 5.55 Control effectiveness for 75 kg [18]

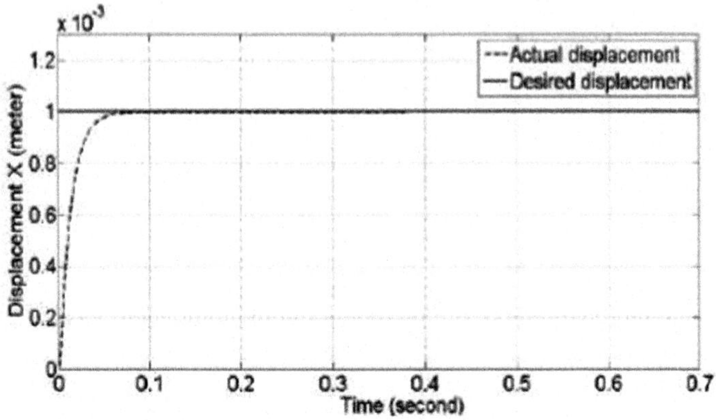

(d) System response $x(t)$ (m = 75 kg).

Fig. 5.55 (continued)

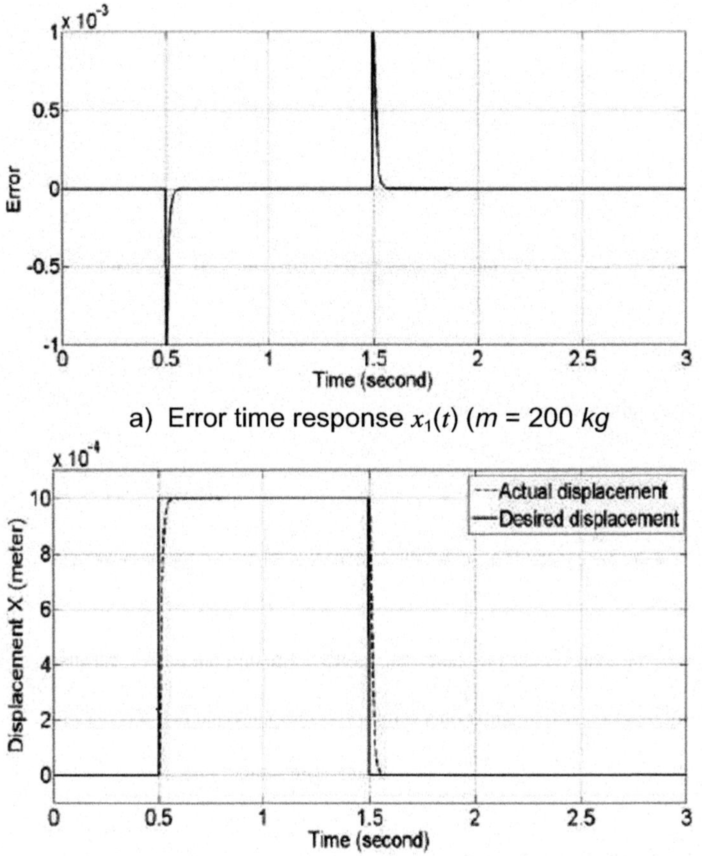

a) Error time response $x_1(t)$ (m = 200 kg

b) Position of system $x(t)$ response (m = 200 kg)

Fig. 5.56 SMC for step position input [18]

5.13.2 SMC Application 2

Reference [19] compares the performance of a first-order linear surface model with a second-order sliding model noting the improvement and a backstepping variable structure control leading to a robust system with exponential decay of error (Figs. 5.57 and 5.58).

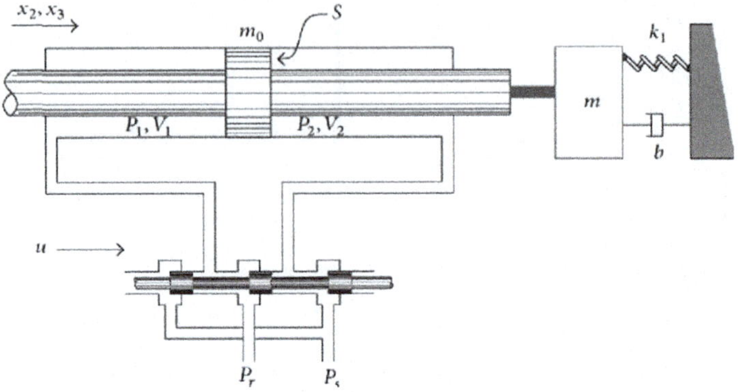

Fig. 5.57 Hydraulic system under study

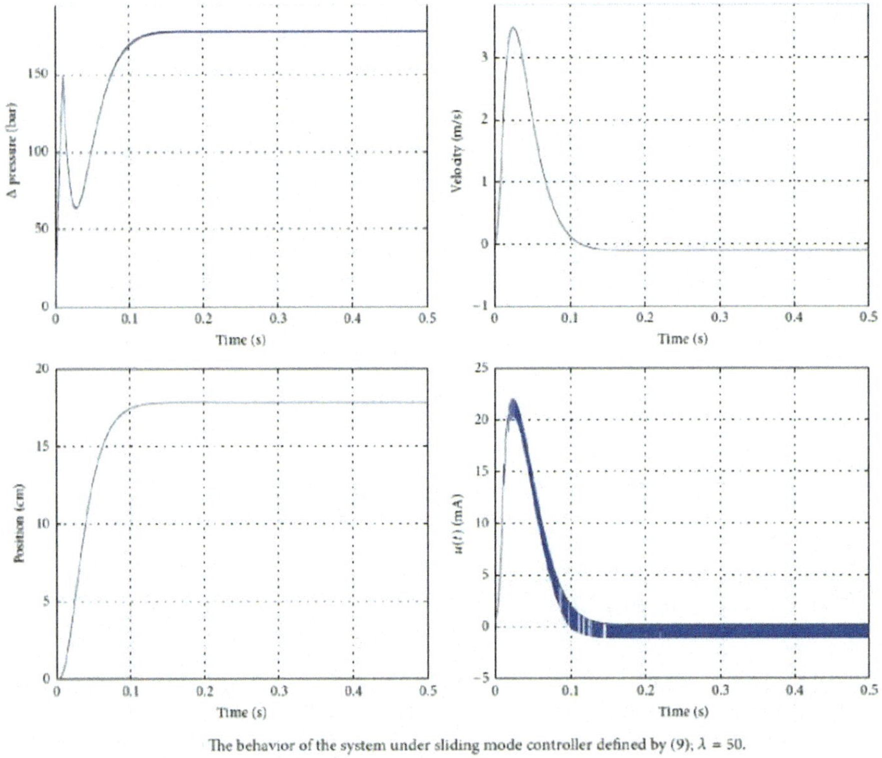

Fig. 5.58 Result with linear surface [19]

Note the steady-state position error (Figs. 5.59 and 5.60).

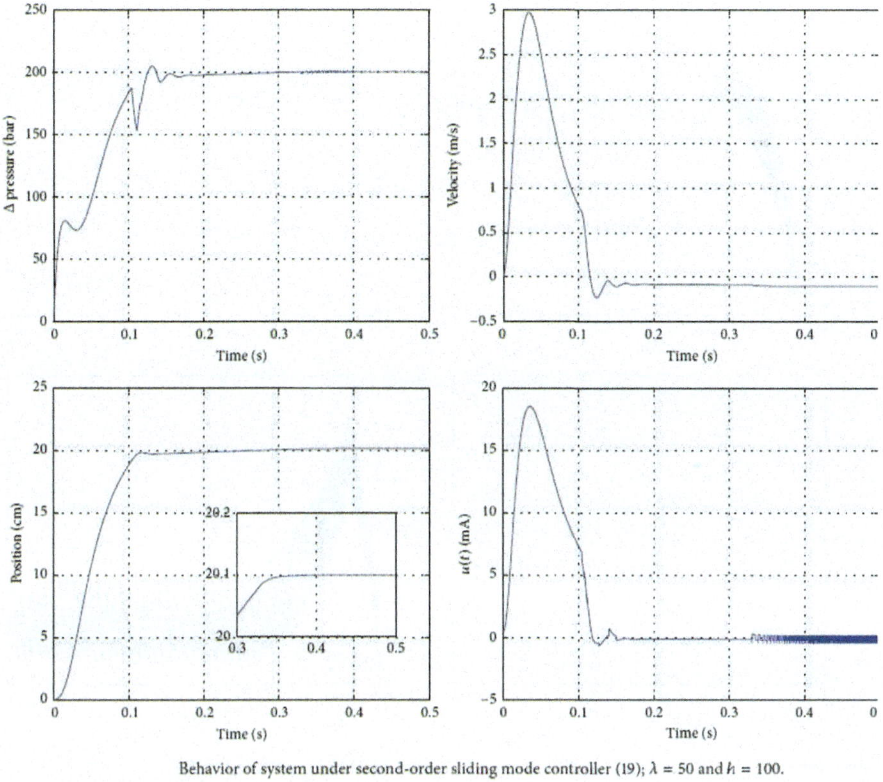

Fig. 5.59 Results with 2nd order sliding mode [19]

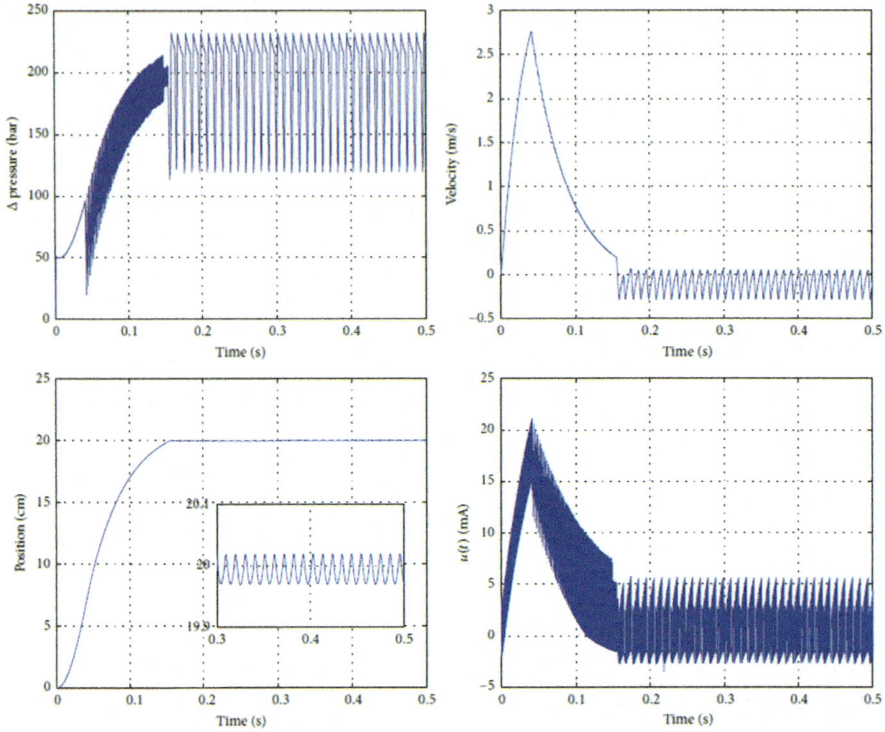

Fig. 5.60 Results of nonlinear surface. Position error in steady state = 0 [19]

Chattering observed, due to the nested sliding mode, was eliminated by treating the discontinuous to a smooth saturation instead of the sigmoid function (Fig. 5.61).

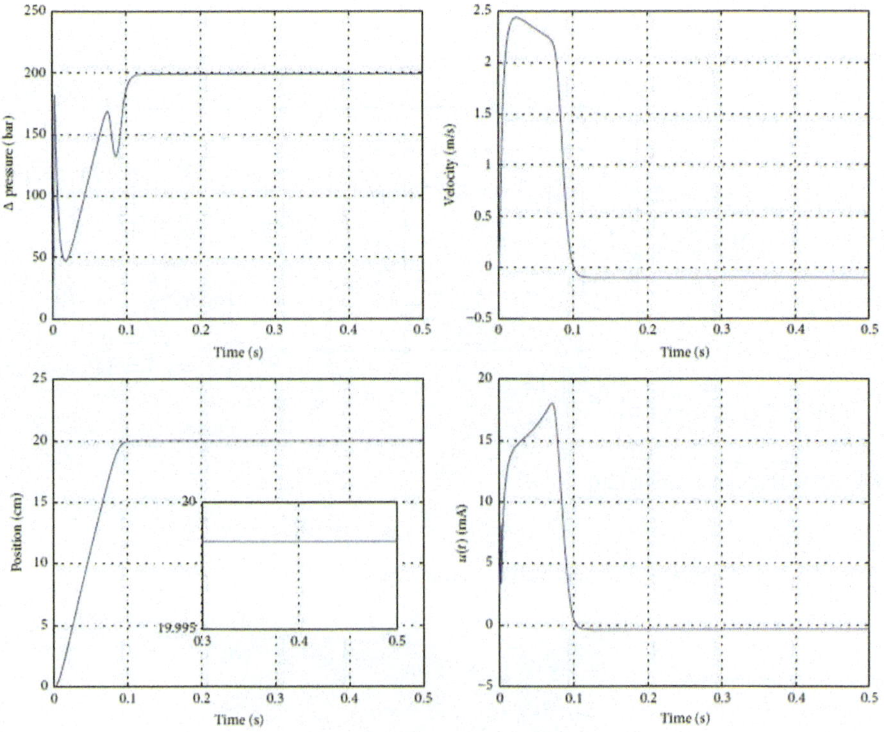

Fig. 5.61 System under sliding mode, nonlinear surface, and saturation [19]

5.13.3 SMC Application 3

Reference [20] describes the application of SMC to the operation of the steam control valve (Figs. 5.62, 5.63 and 5.64).

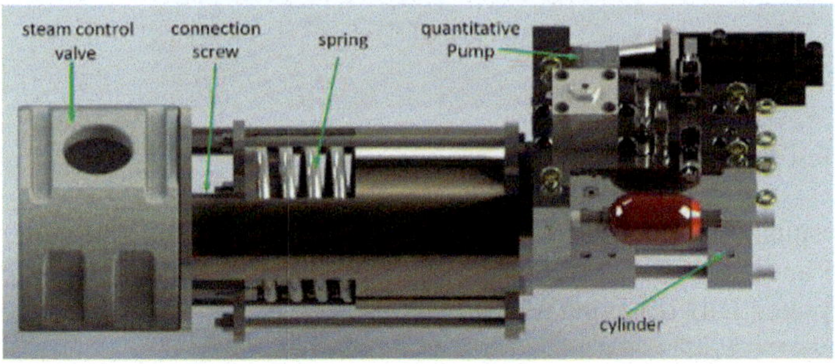

Fig. 5.62 System under study [20]

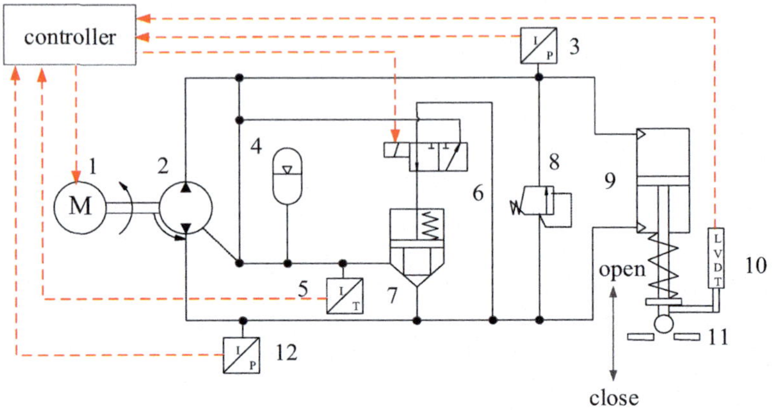

Fig. 5.63 Hydraulic circuit [20]

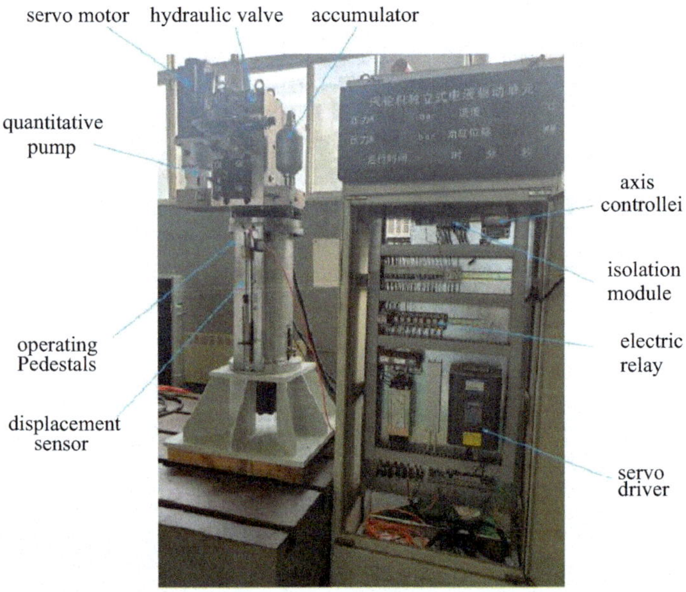

Fig. 5.64 Test platform [20]

The results of the SMC design during valve closing/opening are shown in Figs. 5.65 and 5.66. The design uses experimentally measured friction parameters and an exponentially decaying control law.

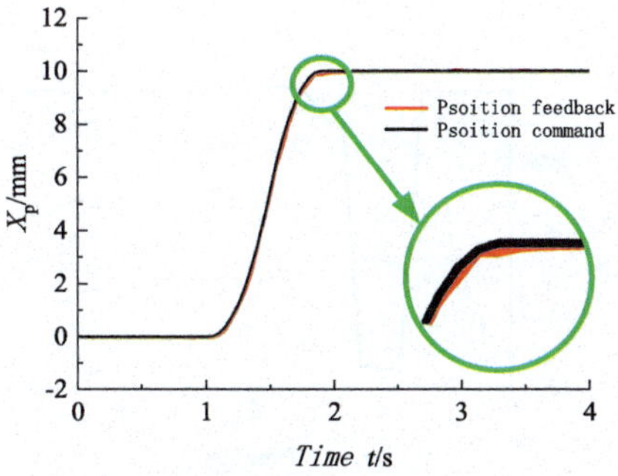

(a) Cylinder displacement curve

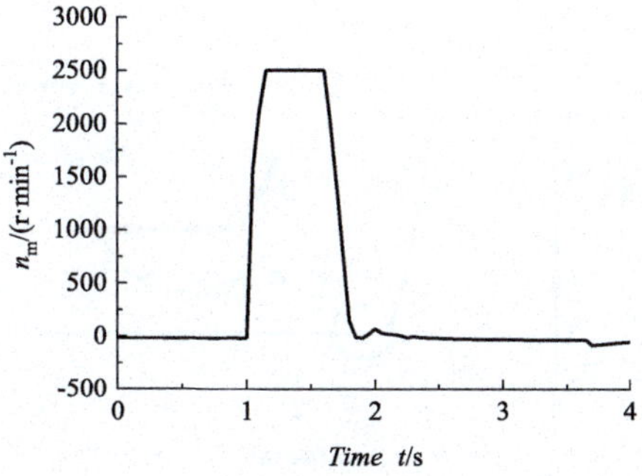

(b) Servomotor speed curve

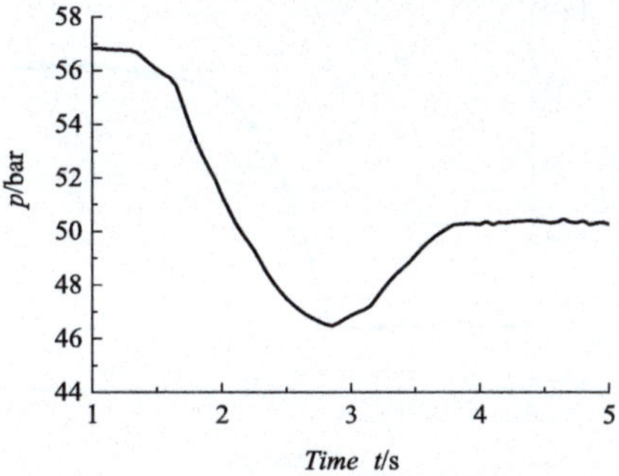

(c) Pressure curve of the high - pressure chamber

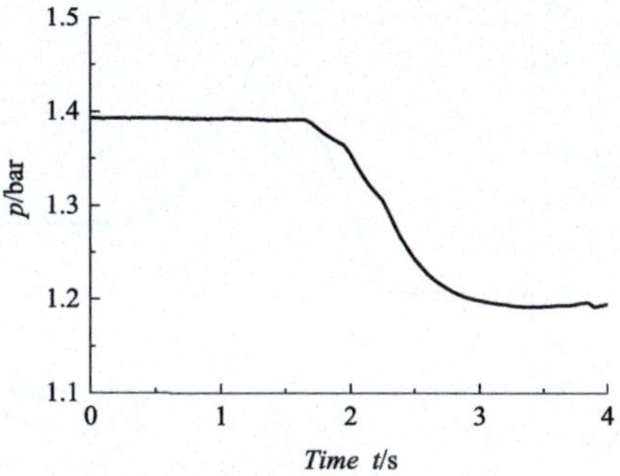

(d) Pressure curve of the low - pressure chamber

Fig. 5.65 Performance in closing valve [20]

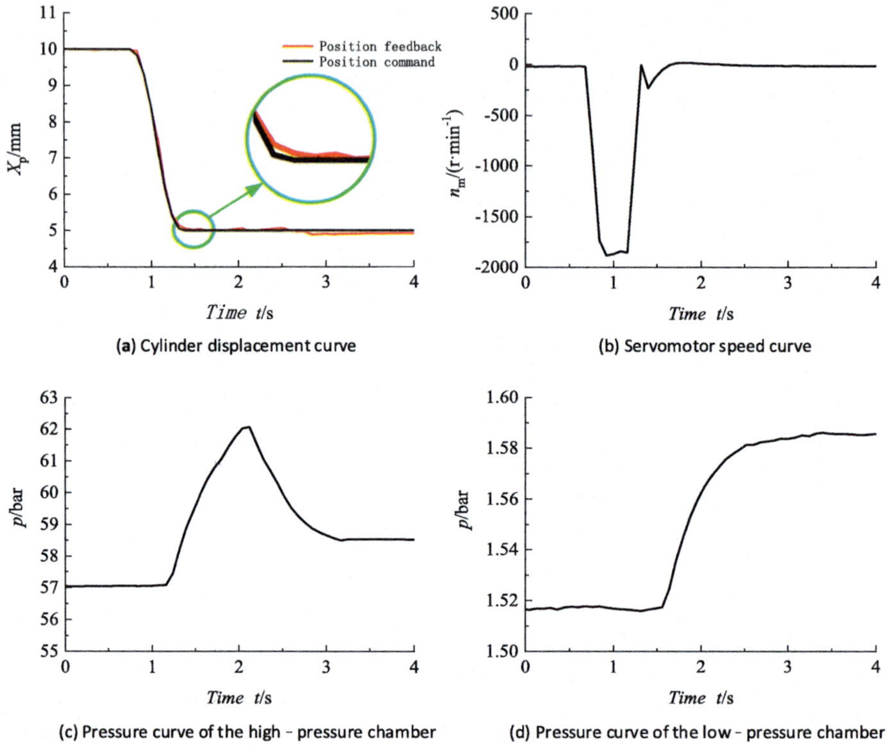

(a) Cylinder displacement curve (b) Servomotor speed curve

(c) Pressure curve of the high - pressure chamber (d) Pressure curve of the low - pressure chamber

Fig. 5.66 Performance in opening valve [20]

In order To alleviate chattering in SMC, the use of a sliding surface with boundary layers weakens chatter at the expense of robustness (Fig. 5.67).

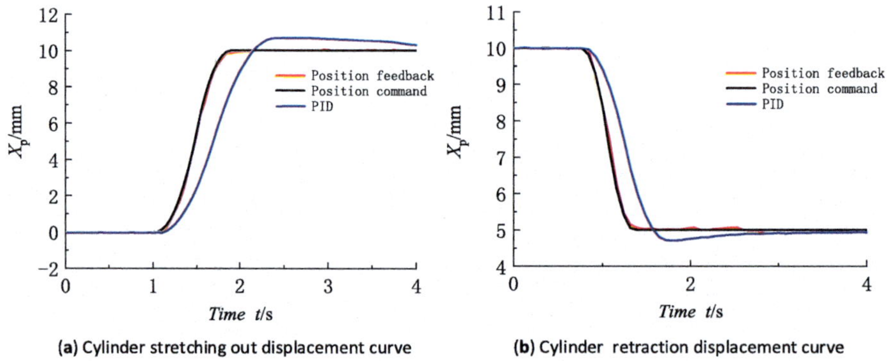

(a) Cylinder stretching out displacement curve (b) Cylinder retraction displacement curve

Fig. 5.67 SMC compared with PID [20]

5.14 Intelligent Control

Intelligent controls based on fuzzy logic, neural networks, and iterative learning algorithms are currently under research. These controls do not need an exact model of the system and are applied with any of the control schemes discussed earlier. A few are cited here.

5.14.1 Fuzzy PID

Intelligent controllers aim to reduce control effort. Reference [21] provides an example of the application of fuzzy PID control to the tracking issue of the EH servo system. Fuzzy logic is applied to regulate the P, I, and D coefficients in control (Figs. 5.68 and 5.69).

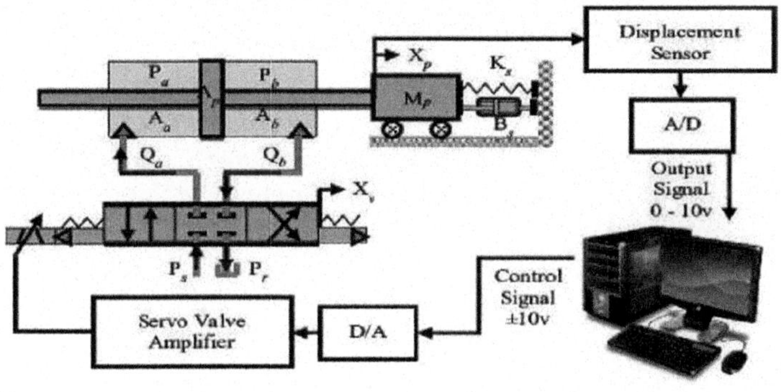

Fig. 5.68 System under study [21]

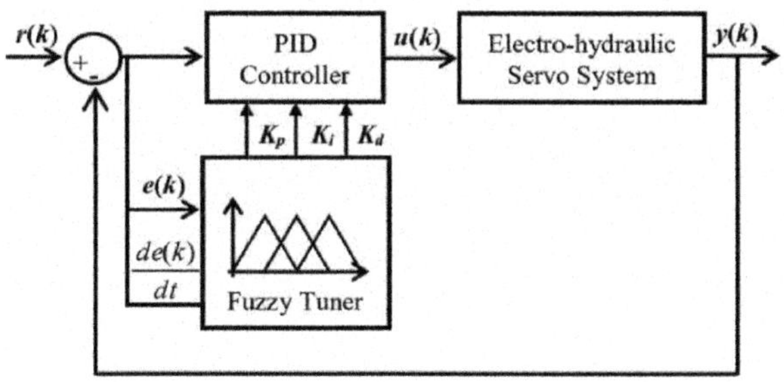

Fig. 5.69 Integration of hybrid fuzzy PID controller [21]

Based on the input error and derivative of error functions, fuzzy rules generate P, I, and D outputs to the system (Figs. 5.70 and 5.71).

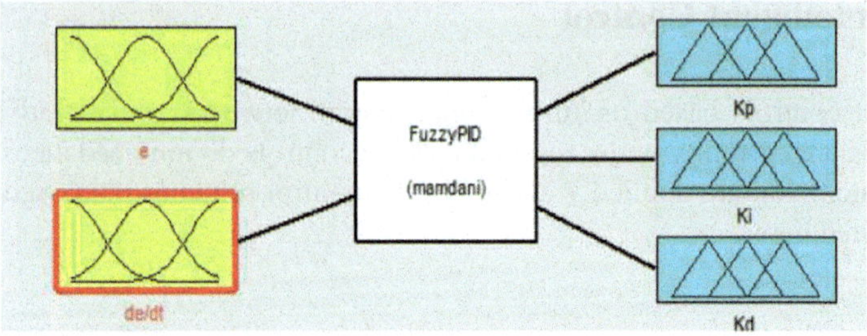

Fig. 5.70 Fuzzy input and outputs [21]

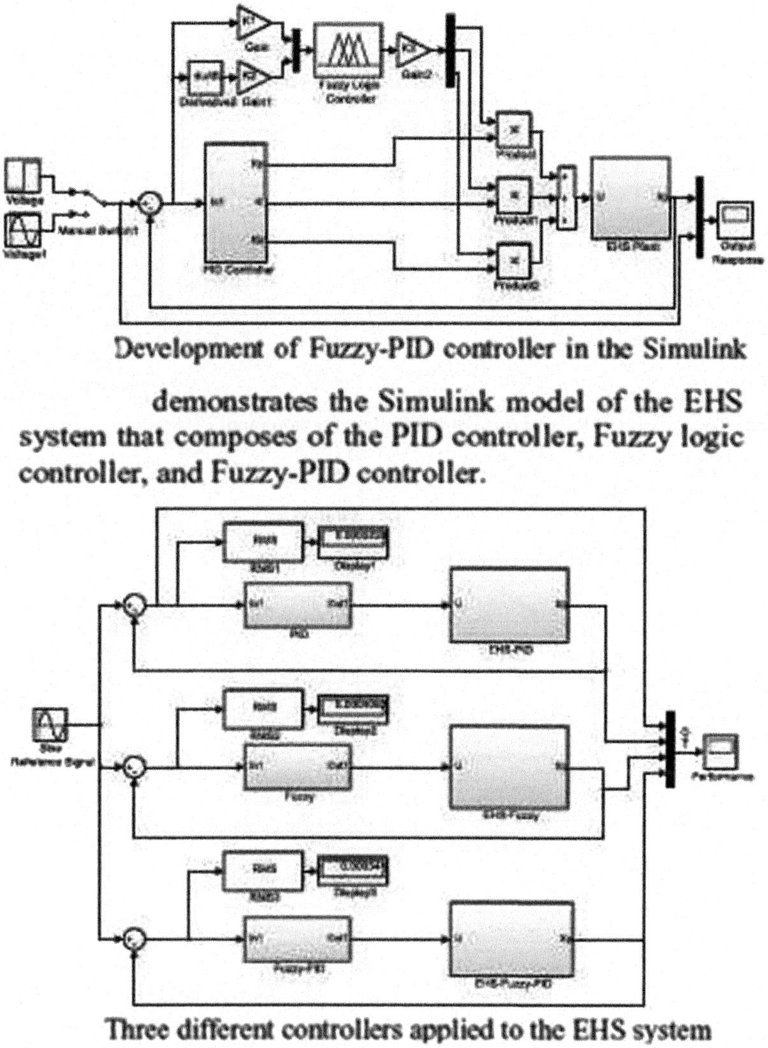

Fig. 5.71 Implementation in simulink [21]

Fuzzy control is facilitated in the Simulink software toolbox. The three control options of PID, fuzzy, and fuzzy PID are compared for step and sine signal inputs (Figs. 5.72, 5.73 and Table 5.6).

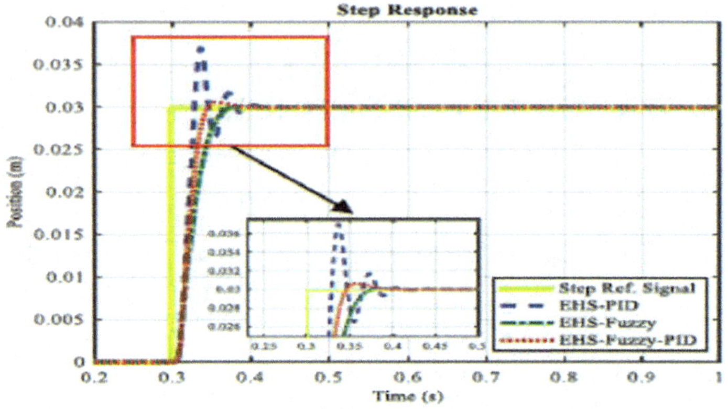

Fig. 5.72 Comparison of the response of controllers for step input [21]

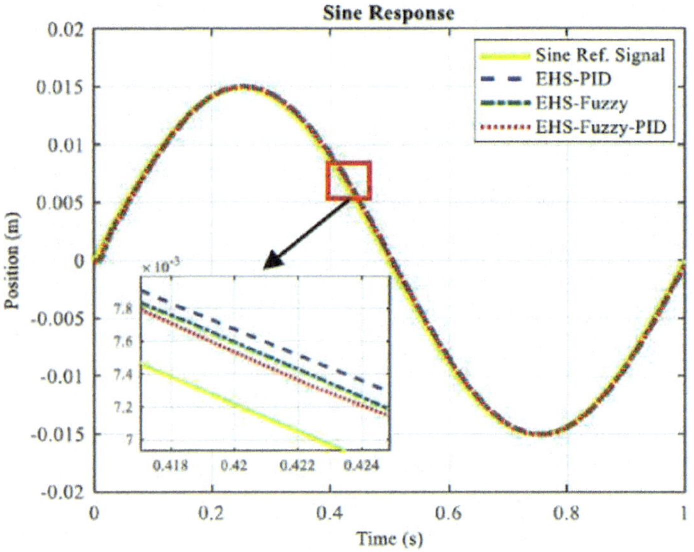

Fig. 5.73 Comparison of the response of controllers for sine input [21]

Table 5.6 Comparison of three controllers [21]

Controller	RMS (step)	RMS (sinusoidal)
PID	7.248×10^{-9}	5.224×10^{-4}
Fuzzy logic	2.099×10^{-11}	4.092×10^{-4}
Fuzzy-PID	9.950×10^{-13}	3.410×10^{-4}

5.14.2 Self-learning Adaptive Fuzzy Controller

The EH system under study includes an unbalanced cylinder servo valve with LVDT feedback for position [22] (Fig. 5.74).

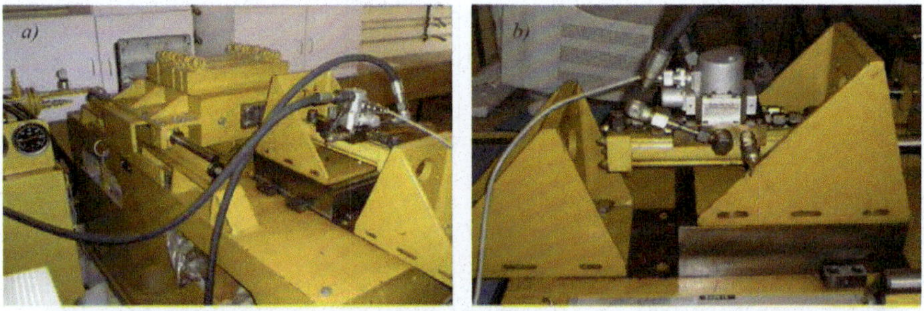

a) Structure of the testing system (cylinder40/28 mm, slider mass 70 to 300 kg, lin. pot. 300 mm ±1%), b) Servo valve (MOOGD-769-233, Q = 19 l/min)

Fig. 5.74 System under study [22]

A combination of feedforward, adaptive fuzzy control, and a switching function is designed as a self-tuning controller to overcome parameter and load variations (Fig. 5.75).

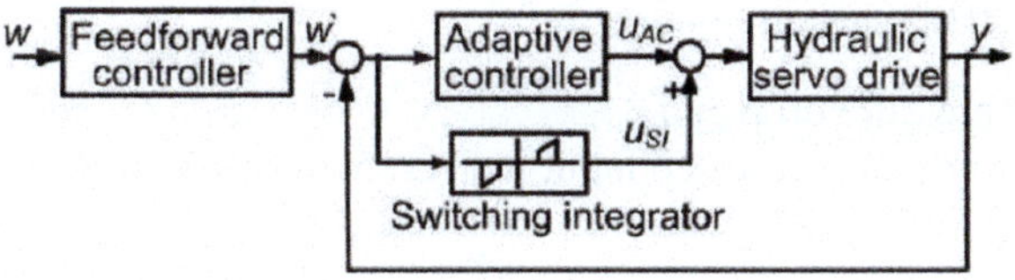

Fig. 5.75 Control structure [22]

With the linguistic variables of error and error derivative as input to the system, fuzzy rules are prepared, and a self-tuning ability is built in with iterations (Fig. 5.76).

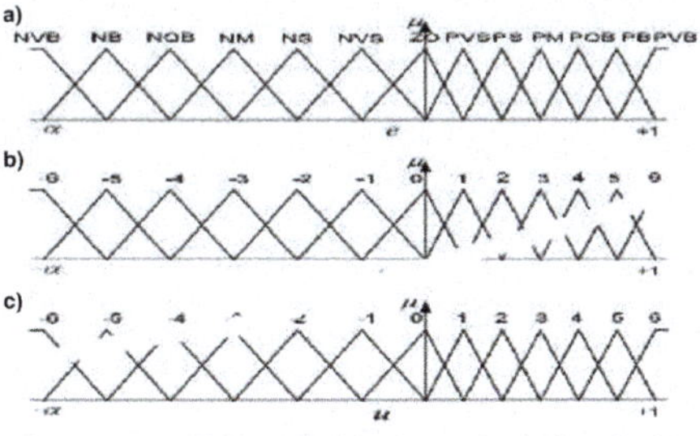

Linguistic variables represented as fuzzy variables; a) e, b) ė, c) u

$\dot{e}(k)$ \ $e(k)$	-6	-5	-4	-3	-2	-1	0	1	2	3	4	5	6
-6	-6	-6	-6	-6	-6	-5	-4	-3	-2	-1	0	0	0
-5	-6	-6	-6	-6	-5	-4	-3	-2	-1	0	0	0	0
-4	-6	-6	-5	-5	-4	-3	-2	-1	0	0	0	0	1
-3	-6	-5	-5	-4	-3	-2	-1	0	0	0	0	1	2
-2	-6	-5	-4	-3	-2	-1	-1	0	0	0	1	2	3
-1	-6	-4	-3	-2	-1	-1	0	0	0	1	2	3	4
0	-6	-4	-3	-2	-1	-1	0	1	1	2	3	4	6
1	-4	-3	-2	-1	0	0	0	1	1	2	3	4	6
2	-3	-2	-1	0	0	0	1	1	2	3	4	5	6
3	-2	-1	0	0	0	0	1	2	3	4	5	5	6
4	-1	0	0	0	0	1	2	3	4	5	5	6	6
5	0	0	0	0	1	2	3	4	5	5	6	6	6
6	0	0	0	1	2	3	4	5	6	6	6	6	6

Rule base or decision table

Fig. 5.76 Self-tuning fuzzy rules [22]

Fuzzy inputs of error and error rates and the logic rules for servo valve input are shown above (Fig. 5.77).

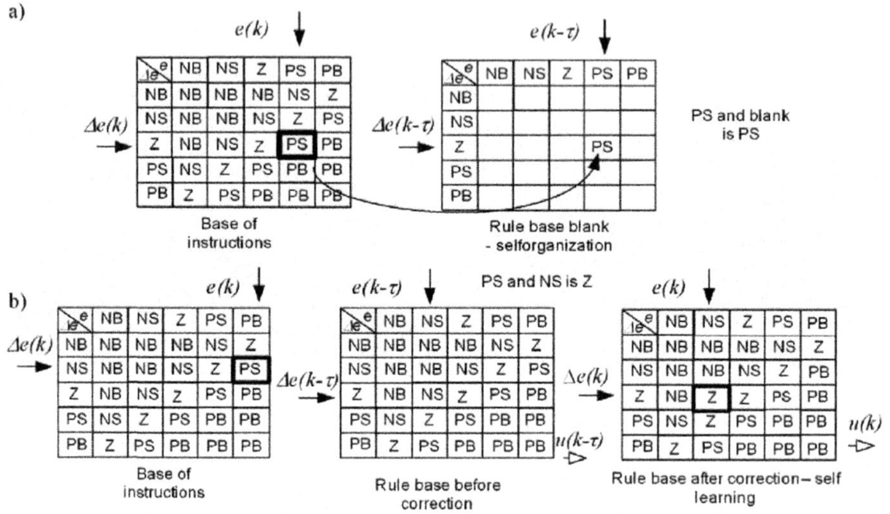

Fig. 5.77 Self-organizing mechanism [22]

The mechanism of self-learning to change rules is incorporated as above. Effectiveness of control is shown for

- 20% supply pressure drop and
- self-learning process for a step input in Figs. 5.78 and 5.79 (Fig. 5.80).

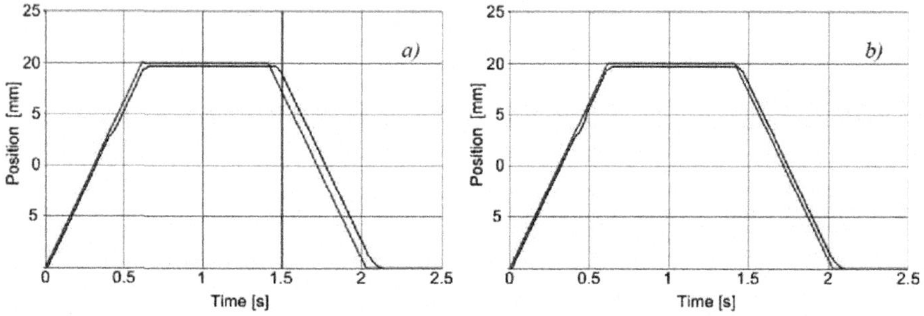

Fig. 5.78 Ramp response [22]

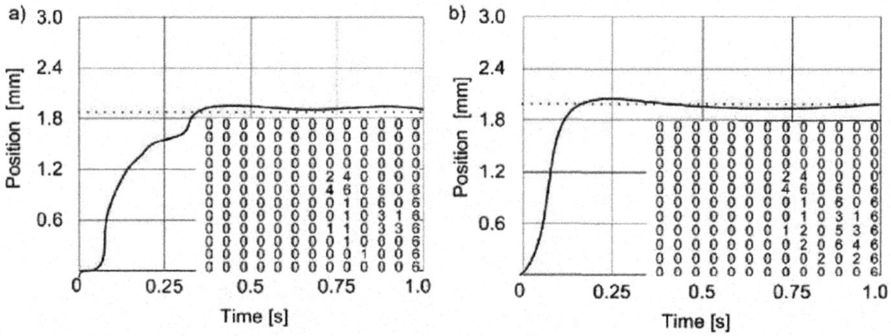

Fig. 5.79 Self learning process for step input [22]

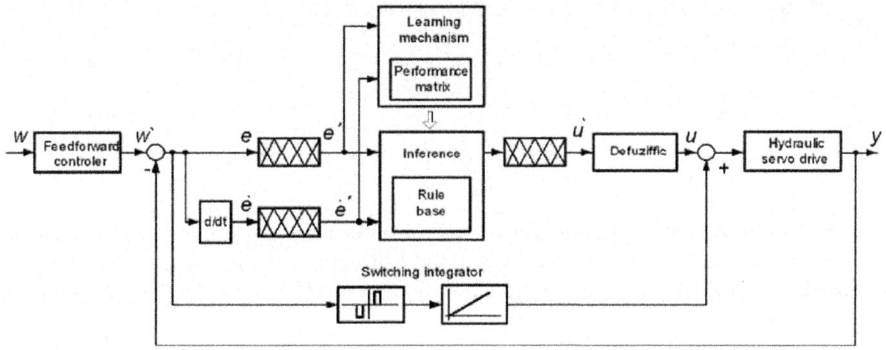

Fig. 5.80 Hybrid controller [22]

The hybrid controller combines the "experienced" fuzzy control with a feedforward controller and the switching integrator. The response of this controller for a ramp input is shown below (Fig. 5.81).

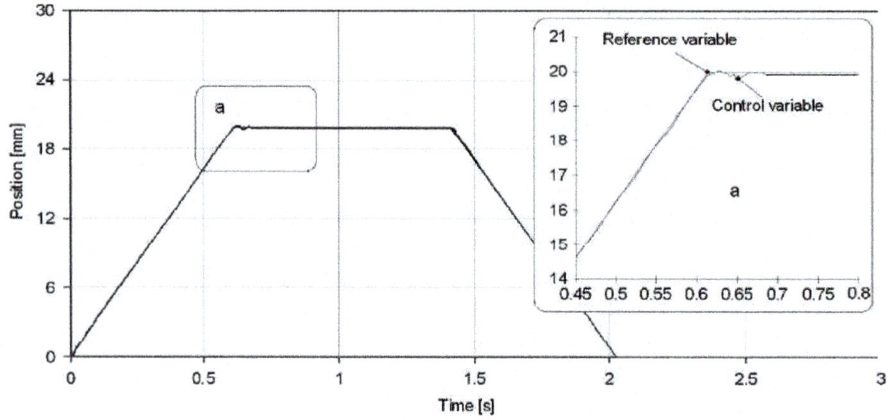

Fig. 5.81 Response of the hybrid controller for ramp input [22]

5.14.3 Fuzzy PID Control of the Hydraulic Crane

Application of fuzzy PID control to a hydraulic crane to improve the performance as well as to reduce vibrations is reported in [23]. Inputs to the fuzzy controller are the error and error derivative functions (Figs. 5.82, 5.83, 5.84, 5.85, 5.86, 5.87 and Table 5.7).

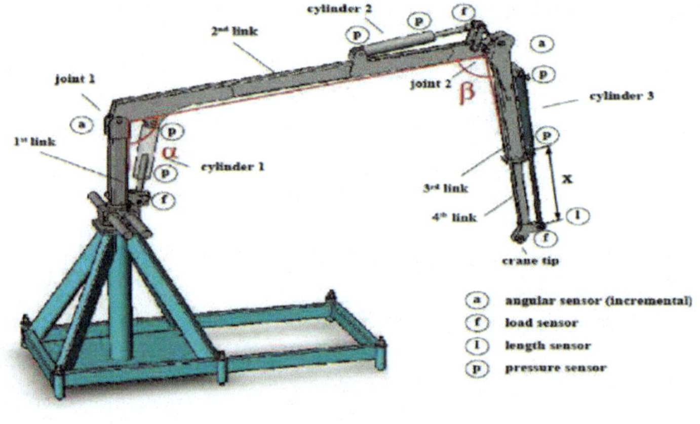

Fig. 5.82 Hydraulic crane [23]

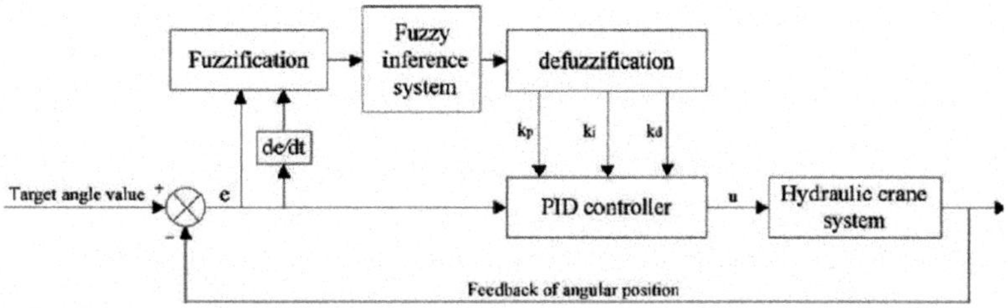

Fig. 5.83 Structure of self tuning fuzzy PID control [23]

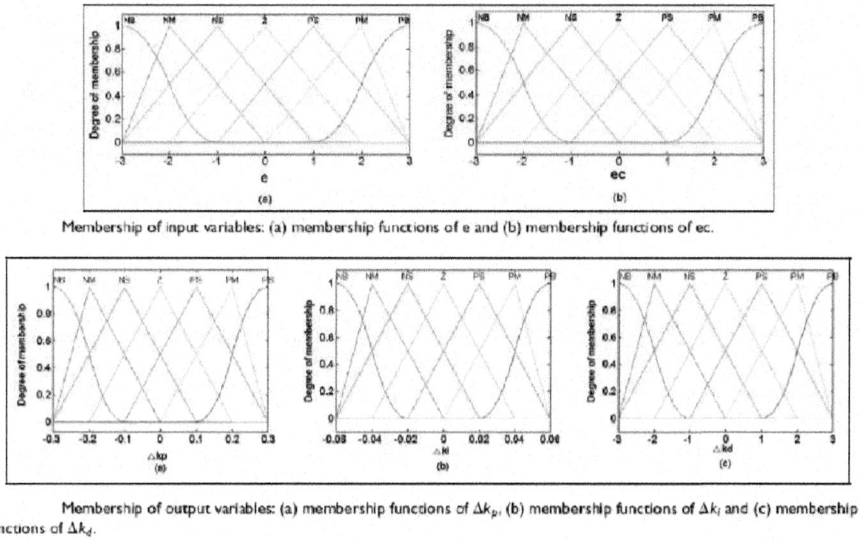

Fig. 5.84 Input and outputs of fuzzy control [23]

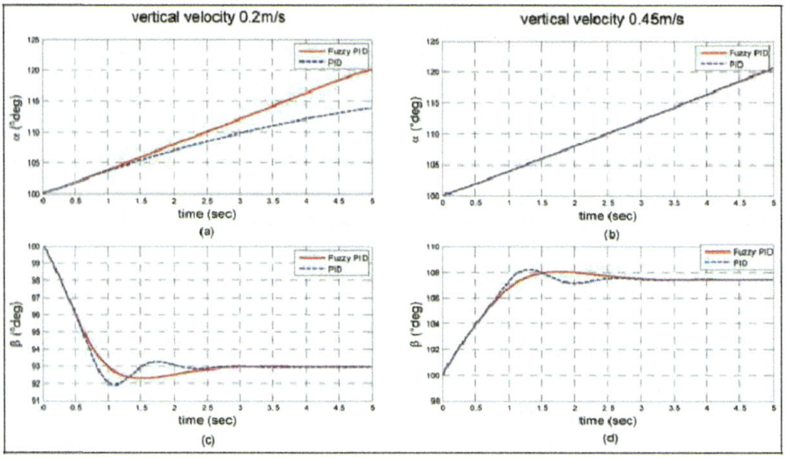

Comparison of joint angular variation: (a) angular variation of alpha at vertical velocity of 0.2 m/s, (b) angular variation of alpha at vertical velocity of 0.45 m/s, (c) angular variation of beta at vertical velocity of 0.2 m/s and (d) angular variation of beta at vertical velocity of 0.45 m/s.

Fig. 5.85 Comparison of PID with fuzzy PID control at two speeds [23]

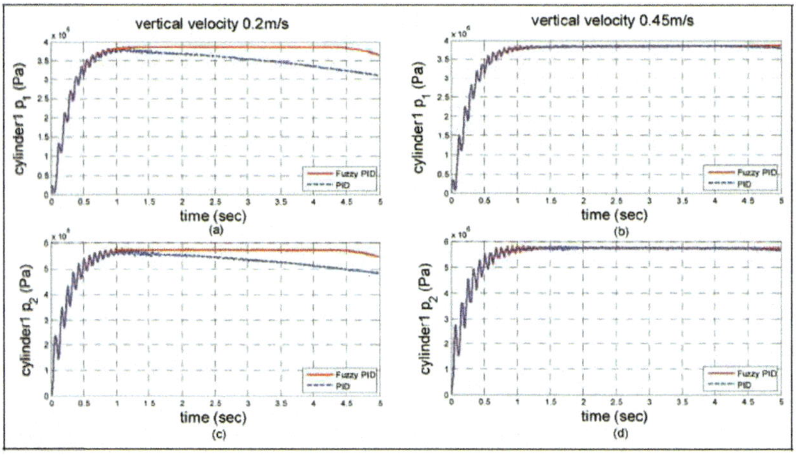

Comparison of pressure in both chambers of cylinder I: (a) pressure of head end chamber of cylinder I at vertical velocity of 0.2 m/s, (b) pressure of head end chamber of cylinder I at vertical velocity of 0.45 m/s, (c) pressure of rod end chamber of cylinder I at vertical velocity of 0.2 m/s and (d) pressure of rod end chamber of cylinder I at vertical velocity of 0.45 m/s.

Fig. 5.86 Comparison of cylinder pressures for PID and fuzzy PID controls [23]

Comparison of pressure amplitudes in both chambers of cylinder 2.

Vertical velocity (m/s)	P1 (MPa)			P2 (MPa)		
	PID	Fuzzy PID	Decreasing percentage (%)	PID	Fuzzy PID	Decreasing percentage (%)
0.1	4.71	3.53	25.05	6.20	4.59	25.97
0.2	4.67	3.58	23.34	6.16	4.64	24.68
0.3	4.77	3.49	26.83	6.26	4.56	27.16

PID: proportional–integral–derivative.

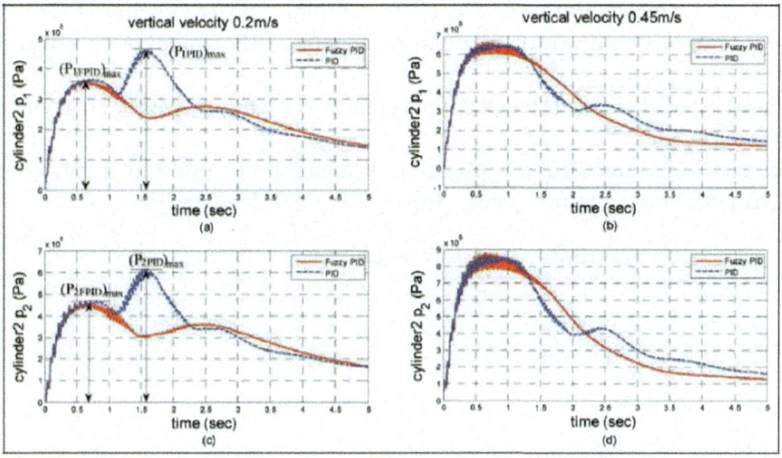

'. Comparison of pressure in both chambers of cylinder 2: (a) pressure of head end chamber of cylinder 2 at vertical velocity of 0.2 m/s, (b) pressure of head end chamber of cylinder 2 at vertical velocity of 0.45 m/s, (c) pressure of rod end chamber of cylinder 2 at vertical velocity of 0.2 m/s and (d) pressure of rod end chamber of cylinder 2 at vertical velocity of 0.45 m/s.

Fig. 5.87 Comparison of cylinder 2 pressures for PID and fuzzy PID controls [23]

Table 5.7 Fuzzy control rules [23]

ec Δk_p/ Δk_i/ Δk_d e	NB	NM	NS	ZO	PS	PM	PB
NB	PB/NB/PS	PB/NB/NS	PM/NM/NB	PM/NM/NB	PS/NS/NB	ZO/ZO/NM	ZO/ZO/PS
NM	PB/NB/PS	PB/NB/NS	PM/NM/NB	PS/NS/NM	PS/NS/NM	ZO/ZO/NS	NS/ZO/ZO
NS	PM/NB/ZO	PM/NM/NS	PM/NS/NM	PS/NS/NM	ZO/ZO/NS	NS/PS/NS	NS/PS/ZO
ZO	PM/NB/ZO	PM/NM/NS	PS/NS/NS	ZO/ZO/NS	NS/PS/NS	NM/PM/NS	NM/PM/ZO
PS	PS/NM/ZO	P5/NS/ZO	ZO/ZO/ZO	NS/PS/ZO	NS/PS/ZO	NM/PM/ZO	NM/PB/ZO
PM	PS/ZO/PB	ZO/ZO/PS	NS/PS/PS	NM/PS/PS	NM/PM/PS	NM/PB/PS	NB/PB/PB
PB	ZO/ZO/PB	ZO/ZO/PM	NM/PS/PM	NM/PM/PM	NM/PM/PS	NB/PB/PS	NB/PB/PB

NB Negative big; *NM* Negative middle; *NS* Negative small; *ZO* Zero; *PS* Positive small; *PM* Positive middle; *PB* Positive big

5.14.4 Artificial Neural Networks

ANNs address problems of nonlinearity and optimization issues of systems. They have an input layer receiving system inputs and pass to a hidden layer with weighting factors that are identified during extensive training over the operating range. The output of the hidden layer is passed on to the output layer of the network. Two basic topologies of ANN are *Back Passing* (BP) and *Radial Basis Function* (RBF). Neural networks can be deployed for PID tuning, adaptive PID tuning, or as a disturbance observer.

Neural network controls receive inputs of error and derivatives of error and output the PID parameters.

Reference [24] details the application of a neural network-tuned PID controller of a hydraulic force control system (Figs. 5.88, 5.89, 5.90 and 5.91).

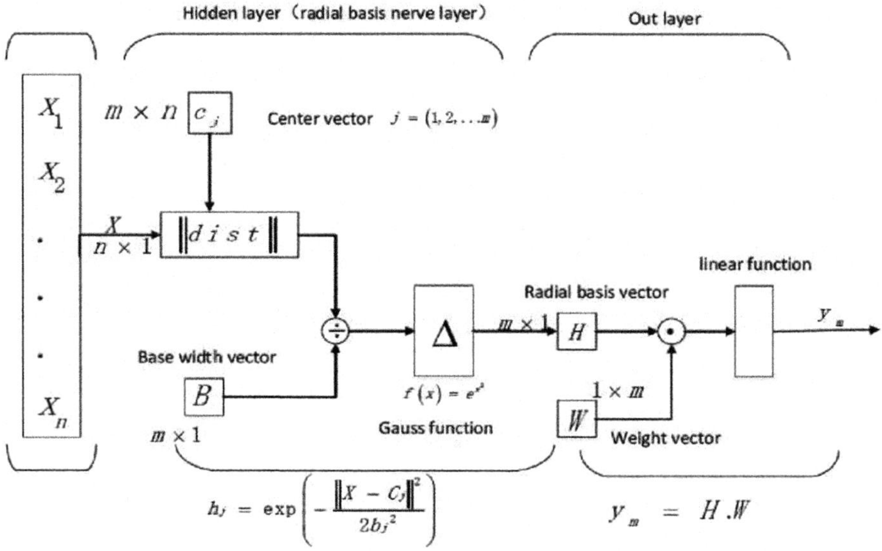

Fig. 5.88 Structure of a neural network [24]

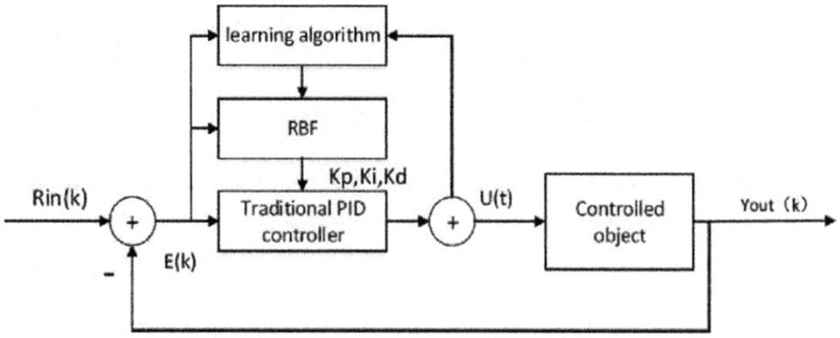

Fig. 5.89 RBF adjust block diagram of PID [24]

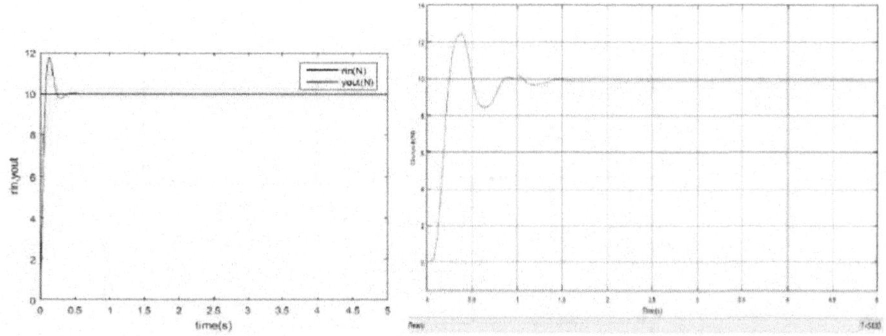

With ANN control Without ANN, PID control

Fig. 5.90 Comparison of performance for step input [24]

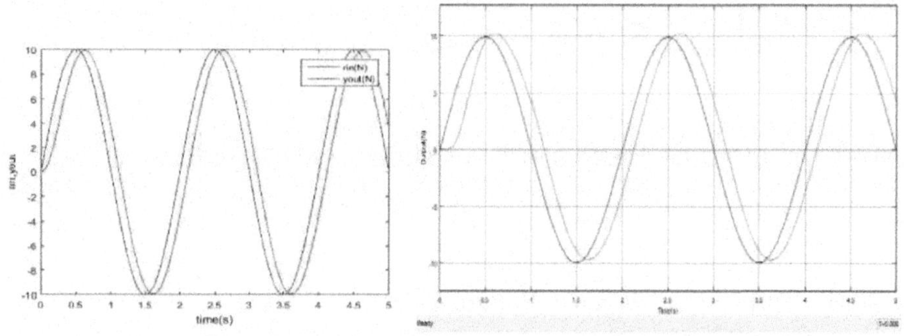

With ANN control PID control

Fig. 5.91 Comparison of performance for sine input [24]

5.14.5 ANN for Active Suspension of Automobiles

Reference [25] describes the application of ANN in the control of an active suspension system of an automobile (Fig. 5.92).

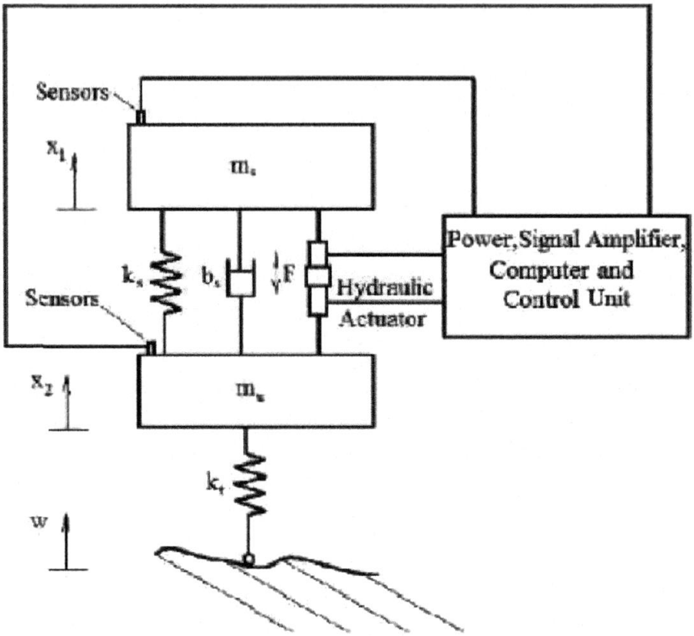

Fig. 5.92 Active suspension model. Simplified quarter car model [25]

5.14.6 *Intelligent Neural Network*

Reference [26] compares the performance of PID, fuzzy PID, and neural network control on an electric servo motor driving a hydraulic pump loaded by a relief valve.

Various responses of controls for step and sinusoidal loading are shown in Figs. 5.93 and 5.94.

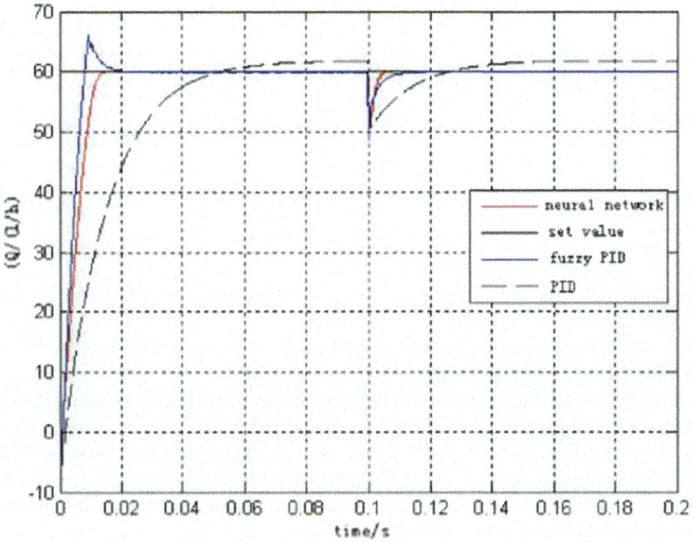

Fig. 5.93 Responses for step loading [26]

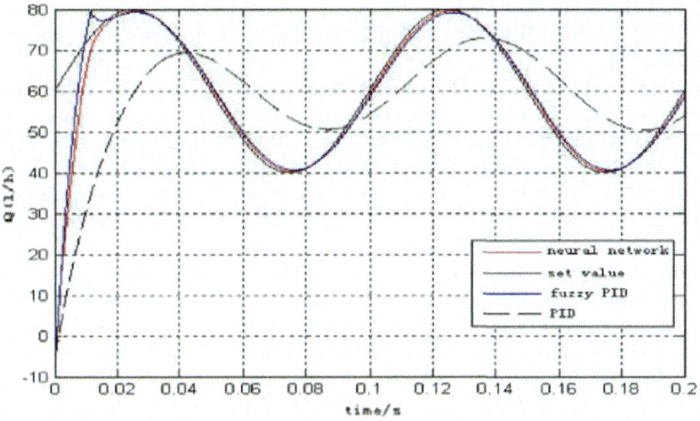

Fig. 5.94 Responses for sinusoidal loading [26]

5.14.7 Iterative Learning Control (ILC)

ILC has the advantage of dealing with uncertain dynamics with little or no prior knowledge. System parameters need not be identified.

Reference [2] describes the application of three types of ILC methods, viz. PID-type filtering in the frequency domain, adaptive ILC, and inverse model ILC for both position- and force-controlled EH servo systems. The relative merits are compared with respect to the root mean square of errors for sinusoidal inputs in both position and force control EH servo systems [2].

References

1. Merritt HE (1967) Hydraulic control systems. Wiley, Hoboken
2. Li L (2015) Simulation and control of servo hydraulic actuators for test applications. Doctoral thesis. Graz University
3. Plummer AR (2007) Control techniques for structural testing: a review. Proc Mech Eng Part 1 J Syst Control Eng 221(2):139–169
4. Valieu S (2015) Nonlinear model and control of electrohydraulic systems. Doctoral thesis. University of Bergamo
5. Shen G, Zhu Z, Li X, Li G, Tang Y, Liu S (2016) Experimental evaluation of acceleration waveform replication on electro hydraulic shaking table: a review. Int J Adv Robot Syst 1–25
6. Gunnarsson S, Krus P (1993) LQG control of hydraulic actuator with flexible mechanical load. In: Triennial world conference, Sydney
7. Mitov A, Kralev J, Slavov T (2020) Comparison of model predictive control (MPC) and linear quadratic Gaussian (LQG) control algorithm for electrohydraulic steering control system. In: E3S web of conferences
8. Kaddissi C, Kenné J-P, Saad M (2004) Position control of an EH servo system. In: Proceedings of 1st international conference on informatics in control, automation and robotics, pp 270–276
9. Feng LJ, Yan H (2020) Nonlinear adaptive robust control of electro hydraulic systems. Appl Sci 10(13):4494
10. Henriksen LC, Poulsen NK (2009) Energy based nonlinear control of hydraulically operated pitch control system. In: EWEL 2009. Proceedings online. Technical University of Denmark
11. Bashetty S, Guillamon JL, Mutnuri SS, Ozcelik S (2020) Design of a robust adaptive controller for the pitch and torque control of wind turbines. Energies 13(5):1195
12. Wos P, Dindorf R (2019) Self tuning controller based on polynomial methods for electro hydraulic servo drives. In: AIP conference proceedings, vol 2077, p 020062. Published online 21 Feb 2019
13. Clegg AC, Dunnigan MW, Lane DM (2000) Self tuning position and force control of an underwater hydraulic manipulator. Ph.D. thesis. Heriot Watt University
14. Frank SA (2018) Control theory tutorial. Basic concepts illustrated by software examples. Springer briefs in applied sciences and technology. University of California
15. Mitov A, Slavov T, Kralev J (2021) Robustness analysis of an electro hydraulic steering control system based on estimated uncertainty model. Information 12:512
16. Wang W, Wang B (2017) Disturbance observer based nonlinear energy saving control strategy for electro hydraulic servo systems. Adv Mech Eng 9(5):1–11
17. Nguyen MH et al (2021) Active disturbance control for position tracking of electro hydraulic servo systems under modeling uncertainties and external load. Actuators 10-00020
18. AL-Samarriaie SA et al (2015) Sliding mode control for electro hydraulic servo systems. IJCCE 15(3)
19. Kosi-glaira E et al (2015) Observer based robust position control of servo systems using variable sliding mode control. Article ID 72495. HINDAWI
20. Yan G, Jin Z et al. Position control study on pump controlled servo motor for steam controlled valve
21. Saeed M, Soon CC, Rozaini G, Ahmad Y, Yahaya MdS, Chai MS (2018) An electro hydraulic servo with intelligent control strategy. In: MATEC web of conferences, vol 150
22. Deticek E, Zuperl U (2011) An intelligent electro hydraulic positioning system. J Mech Eng 5:394–404. http://doi.org/10.5545/sv-june2010.081

23. Hao W, Kan J (2016) Application of self tuning fuzzy PID control of hydraulic crane control system. Adv Mech Eng 8(6)
24. Fang J-W, Li J (2016) A control method of force loading based on BRF. In: Proceedings of 2nd annual conference on advanced material engineering. AME, pp 871–876
25. Pedro JD, Dalvit Ingi OA (2011) Neural network based feedback linearisation control of an electrohydraulic vehicle suspension system. Int J Appl Math Comput Sci 21(1)
26. Yu M (2015) Intelligent neural network control strategy of hydraulic system driven by servo motor. Int J Smart Sens Intell Syst 8(2)

Chapter 6
Design of MIMO EH Servo Systems

Abstract Design methods for MIMO systems with Parallel and Serial Kinematics are presented with case studies. An example of a Serial Kinematic 3 DOF Manipulator with Virtual Decomposition Control is tabled. The design employs SMISMO (IM) valves and the energy savings claimed.

6.1 General

A wide range of hydraulic servo systems can be broadly classified, kinematically, into serial jointed or parallel joints. Hydraulic cranes in handling forestry or offshore applications fall into serial robotic kinematic structures, while motion simulators have parallel kinematics. The design of the power elements of the individual joints/legs is based on the load locus (force/velocity) needs of the drive. While the stroke/velocity drive requirements are calculated from the inverse kinematics of Cartesian movements of load, the force estimates for design are read from the inverse dynamics of load.

Two case studies, one of each type, are discussed here.

6.2 Parallel Kinematics

Specifications for a 6 DOF Stewart platform to simulate sea state conditions are noted in Table 6.1.

D. R. Raghavendra, *Electrohydraulic Servo Systems*, Springer Tracts in Mechanical Engineering, https://doi.org/10.1007/978-981-19-8065-7_6

Table 6.1 Specifications of a 6 DOF simulator [1]

Motion	Independent (M) or (°)	Superimposistion (M) or (°)	Peak velocity (M) or (°)	Peak acceleration (M) or (°)	Frequency
HEAVE	± 0.60	± 0.30	1.5	Max. dynamic payload not to exceed 300kN vertical and 50kN lateral	Actuator velocity not to exceed 1.5 m/s. Power demand not to exceed rated peak power(1 MW)
SURGE	**± 0.30**	**± 0.15**	**1.5**		
SWAY	**± 0.30**	**± 0.15**	**1.5**		
PITCH	**± 3**	**± 1.5**	**10**		
ROLL	**± 3**	**± 1.5**	**10**		
YAW	**± 3**	**± 1.5**	**10**		
ACTUATOR	–	–	**1.5**		

For the specified work envelope, the size of the fixed base and moveable platform is 5.7 m × 5.7 m and the distance/height between them and the individual leg lengths are 6.2 m (Fig. 6.1).

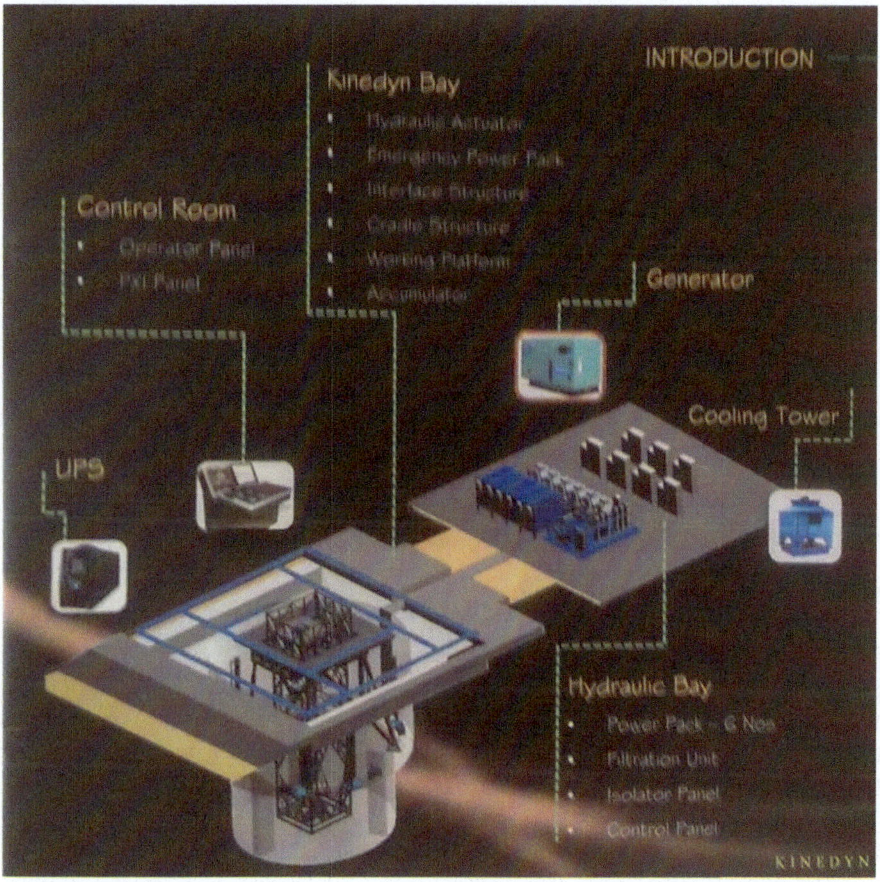

Fig. 6.1 6 DOF sea state simulator [1]

Actuating hydraulic cylinders formed an integral part of the legs. The heave specification of 30 ton mass moving at an acceleration of 1.2 G was clearly the deciding factor for the sizing of the power elements (Fig. 6.2).

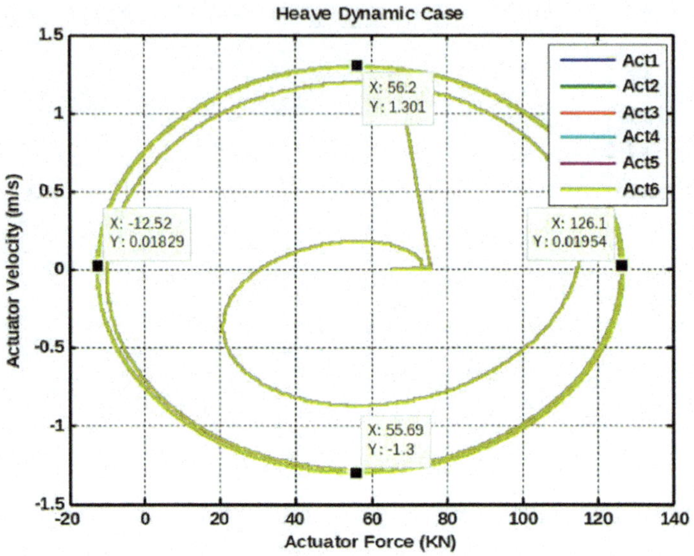

Fig. 6.2 Load locus of platform [1]

A tandem actuator with hydraulic pressure on the piston rod with an optimum-sized accumulator was designed to minimize the control power. The control power was further reduced by use of VCD servo proportional valves rated 200 l pm at 10 bar pressure drop. The design permitted adequate power for other control and cooling requirements with a budgeted 1000 kW power supply (Figs. 6.3, 6.4 and 6.5).

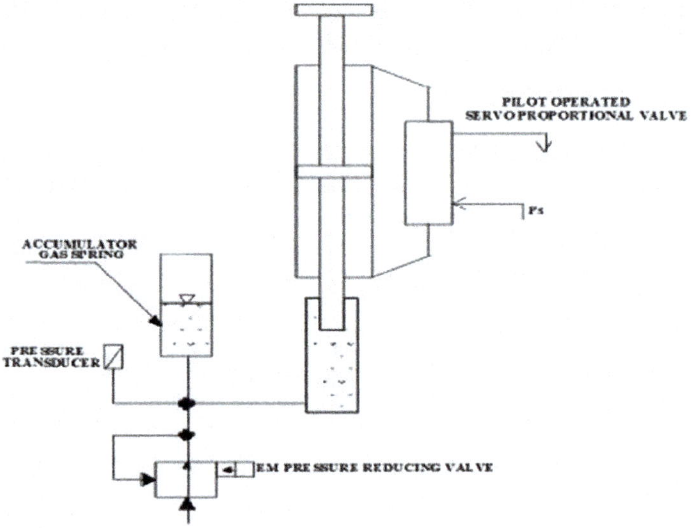

Fig. 6.3 Tandem actuator [1]

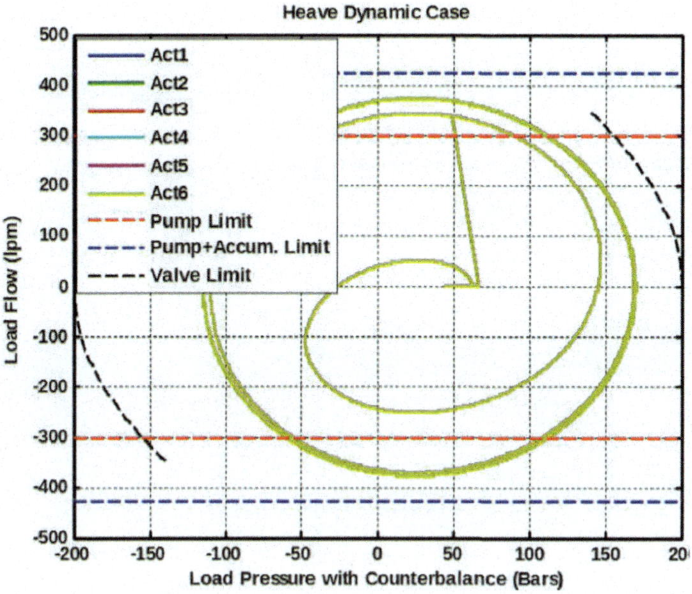

Fig. 6.4 Load locus with counterbalance [1]

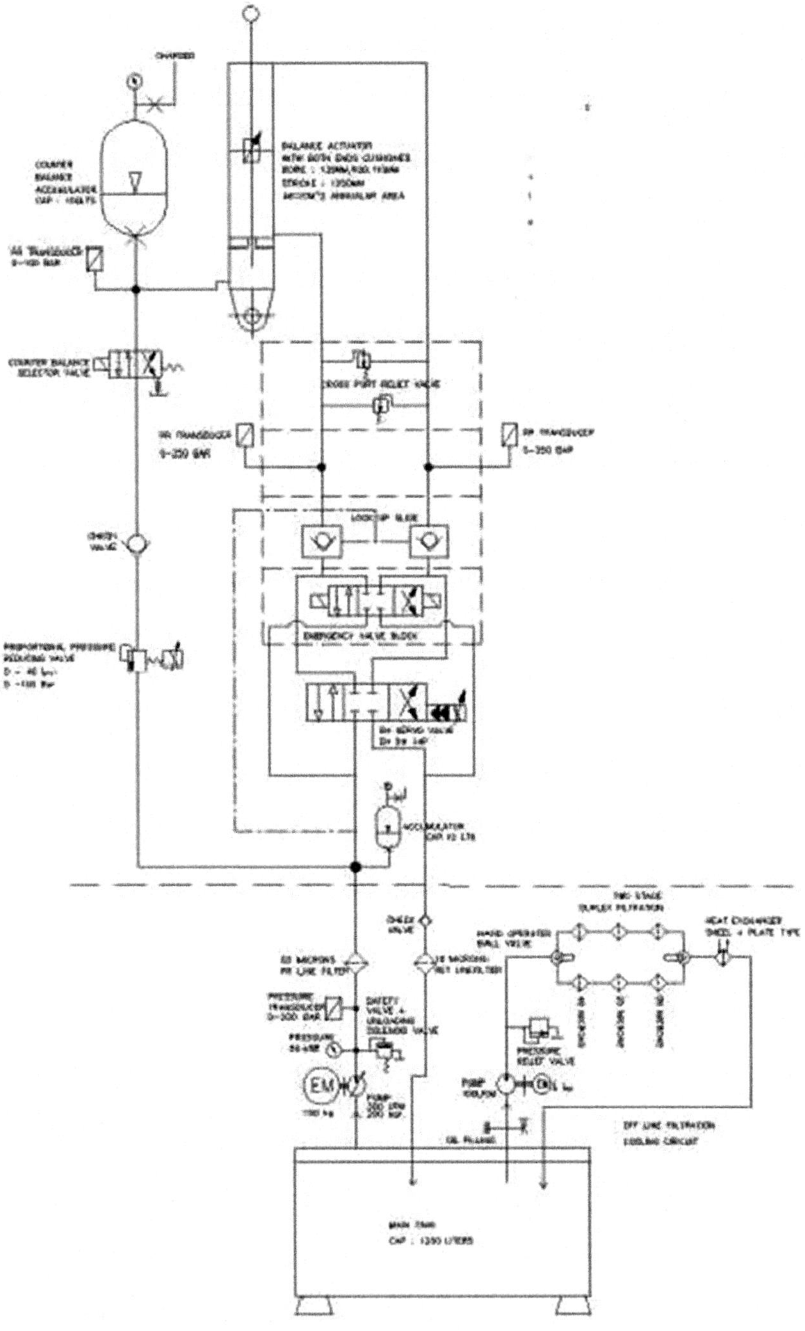

Fig. 6.5 Hydraulic circuit for each actuator [1]

6.3 Serial Kinematics

A forestry crane for handling logs cited as a case study. The specifications include the work envelope, load handling capacity, and system requirements of tracking accuracy and robustness [2] (Figs. 6.6 and 6.7).

Fig. 6.6 Forestry crane [2]

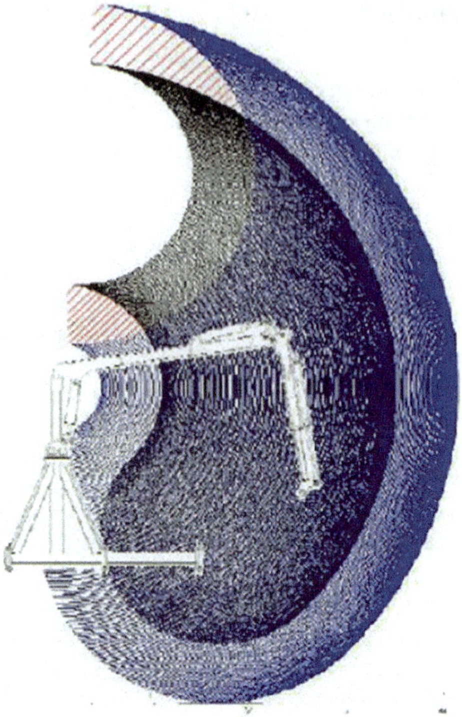

Fig. 6.7 Work envelope [2]

From the inverse kinematics, individual joint movement and speeds are obtained. The actuator center distances, strokes, and actuator speeds are determined by the layout of the design. Actuator center distances/actuator strokes are determined (Fig. 6.8).

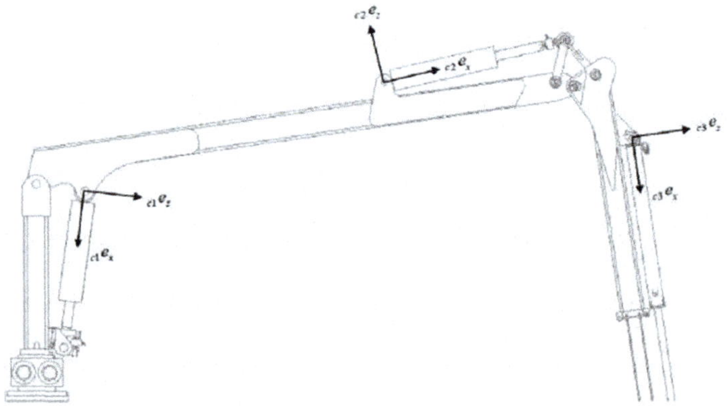

Fig. 6.8 Layout of hydraulic crane [2]

A dynamic model relates load torque, position, velocity, and acceleration to individual torques, position, velocity and acceleration of joints. The joint torques include the friction components of the drive actuators. From the inverse dynamic model of the system, individual joint torques/forces from actuators are derived.

The dynamic model is based on Euler–Newton or Euler–Lagrange formulations. The choice of formulation depends on the number of links in the chain, topology of the chain (series/parallel), position orientation of coordinate frames, and whether recursive procedures are needed.

Euler–Newton is preferred for multiple DOF systems where a recursive structure is needed. It is also possible to compute a dynamic model with Euler–Lagrange with a recursive procedure. The final choice is an individual preference [3]. Here, the Euler–Lagrange formulation is preferred, as there are fewer links in parallel topology. The model is rendered accurate by dynamic parameter identification.

The individual actuator, servo valve rating, and hydraulic power supply are sized according to the load/velocity requirements.

The hydraulic power supply is then sized for the maximum combined flow requirement of actuators for the designed maximum Cartesian speed of the end effector at the designed supply pressure.

The tools required are SolidWorks, MATLAB, and Simulink packages.

6.4 SMISMO-Based Design of the Hydraulic Crane

The design of the hydraulic crane control system based on the SMISMO concept aims to substantially increase the energy efficiency of the system [4] (Fig. 6.9).

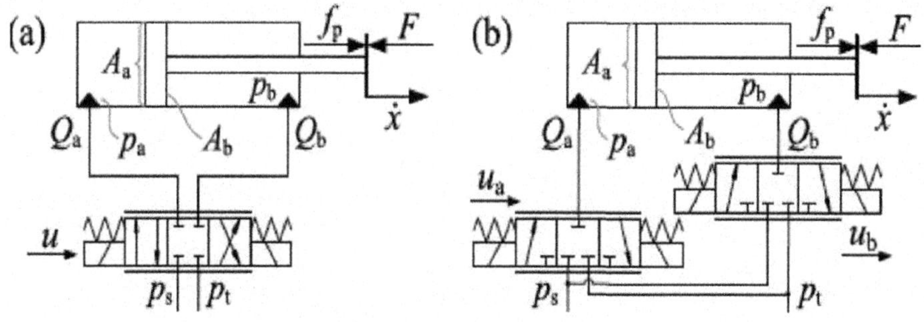

Fig. 6.9 Conventional and SMISMO configurations [4]

Conventional valve-controlled systems control load pressure to overcome force and have no control of individual chamber pressure, resulting in high-supply pressure ($1\frac{1}{2}$ of load pressure). SMISMO controls individual chamber pressures, designs a constant pressure difference **δ** across both the meter-in and meter-out valves.

$$P_s = P_a + \delta P_c = \frac{F_{p\max} + (A_a + A_b)\delta P_c}{Aa}$$

By setting

$$\delta Pc < \left(\frac{Fp \max}{2(A_a + A_b)} \right)$$

The energy saving condition is obtained.

For 100 kN, with A_a= 50.2 cm^2, A_b = 34.4 cm^2, δP_c < 59 bar.

Typically, it is set between tank pressure and 20 bar.

This can also be shown to be valid for operations in other quadrants. Further, energy savings can be realized by controlling Ps (Fig. 6.10).

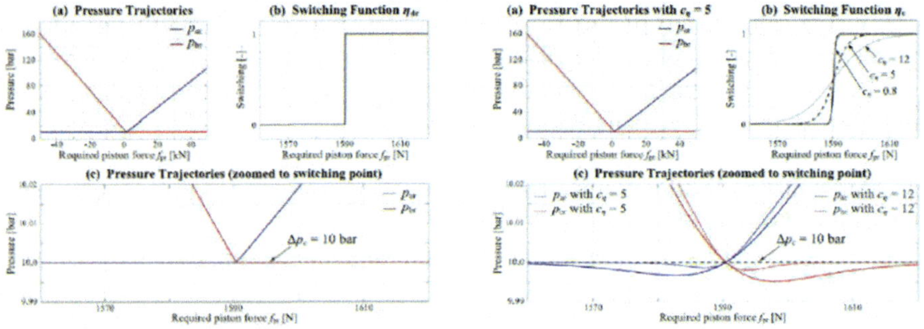

Fig. 6.10 Typical pressure trajectories [4]

The typical trajectories desired show a discontinuous switching function that is approximated by a trigonometric function to enable control. In virtual decomposition control (VDC), the original system is decomposed into modular subsystems based on Newton–Euler dynamics (Fig. 6.11).

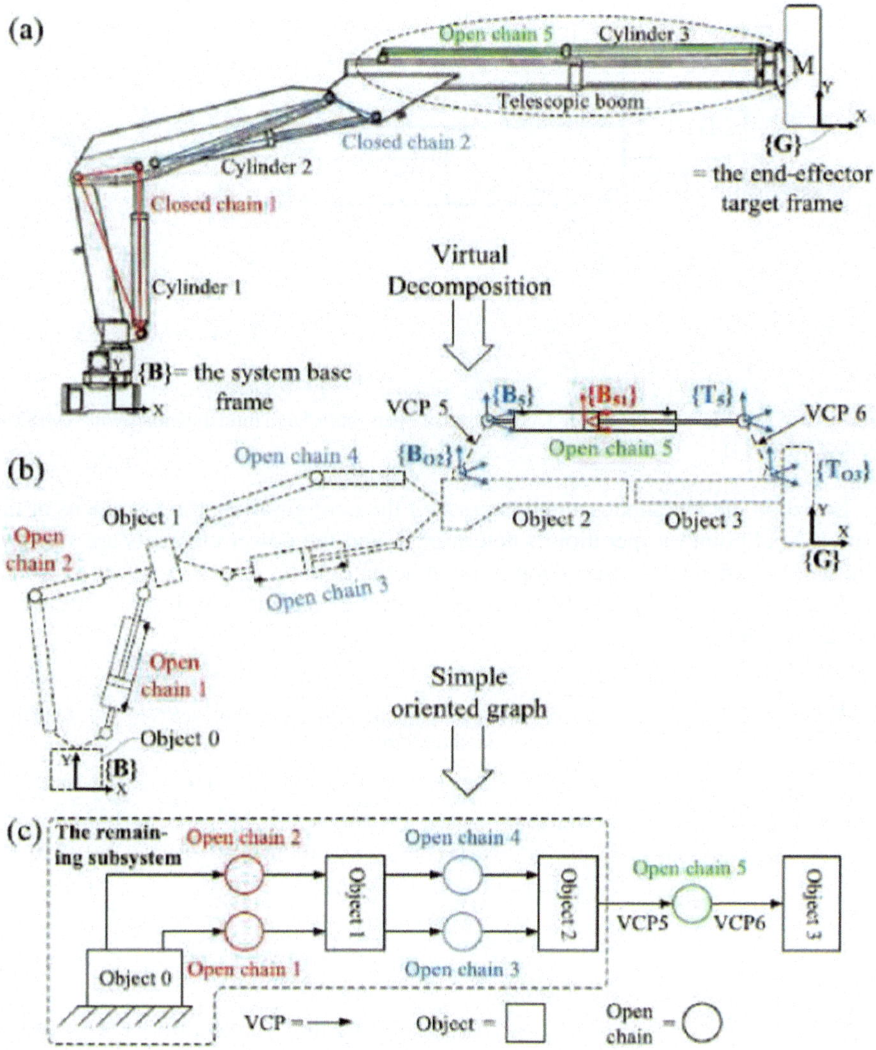

Fig. 6.11 VDC of the 3 DOF manipulator [4]

VDC is performed by placing virtual cutting points (VCPs), which are directed, separated, interfaces that cut through rigid objects. The interface forces/moments are seen as either driving or driven VCPs. Conversion to an open chain reduces to a modular design of individual actuators.

The simple-oriented graph shows source/sink VCPs. (Fig. 6.12)

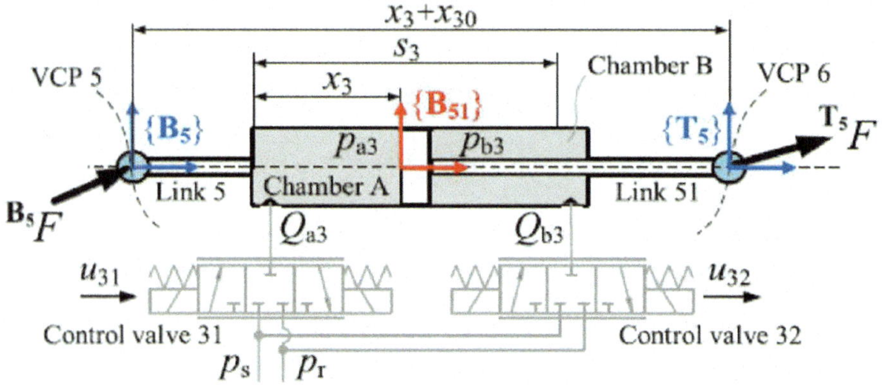

Fig. 6.12 SMISMO out of cylinder 3, composed of open chain 5 and fluid dynamics with SMISMO control layout [4]

Based on the kinematics and dynamics of the configuration, the load locus or the critical load point of operation is determined, and the power elements are sized. A similar exercise leads to the sizing of other actuators.

A comparison of the SMISMO design with the conventional design is shown in Tables 6.2 and 6.3.

Table 6.2 Design comparison [4]

Conventional hydraulic cylinder control; see (76) and (111) in Koivumaki and Mattila (2015a)			
	Cylinder 1	Cylinder 2	Cybnder 3
λ_l[m/s]	30	34	27
$k_{fl}[m^2/(N \cdot s)]$	$6.0 \cdot 10^{-8}$	$8.0 \cdot 10^{-8}$	$20.0 \cdot 10^{-8}$
k_l[m]	0.030	0.030	0.028
The proposed SMEMO control; see (38), (57), and (64)			
	Cybnder 1	Cybnder 2	Cybnder 3
λ_l [m/s]	32	28	26
$k_{pl1}[m^2/(Pa \cdot s)]$ tp	$5 \bullet 10^{-10}$	$3.0 \bullet 10^{-10}$	$5.0 \bullet 10^{-10}$
$k_{pl2}[m^2/(Pa \cdot s)]$	$4 \bullet 10^{-10}$	$1.4 \bullet 10^{-10}$	$2.1 \bullet 10^{-11}$
k_{xl1}[m]	0.020	0.026	0.010
k_{xl2}[m]	0.020	0.005	0.002

Table 6.3 Energy savings [4]

Hydraulic energy consumptions				
	Cyl1 [J]	Cyl2 [J]	Cyl3 [J]	Total [JJ]
Conventional	2.56×10^4	2.10×10^4	1.44×10^3	4.80×10^4
SMISMO	1.64×10^4	0.93×10^4	0.71×10^3	2.64×10^{J}

The integration of energy savings of SMISMO design over time is evident. (Fig. 6.13)

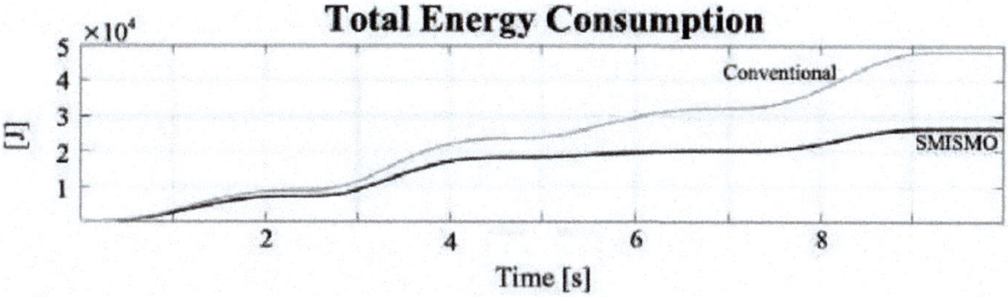

Fig. 6.13 Energy savings with the SMISMO design over time [4]

References

1. Raghavendra DR Kinedyn-design report on control systems
2. Heine A (2008) Modelling, simulation and control of a hydraulic crane. Master's Thesis. Munich University, Diva portal
3. Hoifodt H (2011) Dynamic modelling and simulation of robot manipulators. Master of Science Thesis. Norwegian University of Science and Technology
4. Koivumaki J, Zhu WH et al (2019) Energy efficient and high precision control of hydraulic robots. J Control Eng Practice. Elsevier

Chapter 7
Control Options—MIMO

Abstract The additional challenge of Coupling in MIMO systems are explained and methods for overcoming cited .Centralised and Decentralised Controllers for systems working in free space and in constrained spaces are detailed with examples.Control in constrained spaces demand both position and Force Control. Examples are drawn for both serial, parallel kinematic systems.

7.1 Coupling in MIMO Systems

MIMO systems have an additional challenge of coupling between axes in addition to other challenges of SISO systems.

A 2 DOF system with 2 linear actuators intended to provide vertical and inclination control is shown in Fig. 7.1.

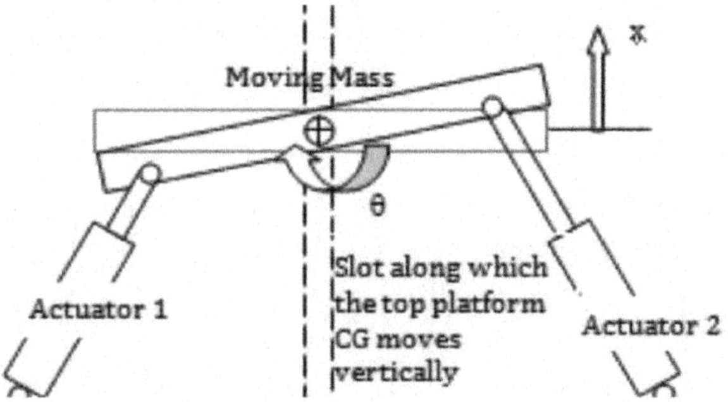

Fig. 7.1 2 DOF system [1]

A ramp input followed by sinusoidal motion to the linear axis is given. Even with a small pitch angle difference, coupling between the two axes comes into play. The system does not settle into the desired response. This uncontrolled system behavior under coupling cannot be accepted (Fig. 7.2).

D. R. Raghavendra, *Electrohydraulic Servo Systems*, Springer Tracts in Mechanical Engineering, https://doi.org/10.1007/978-981-19-8065-7_7

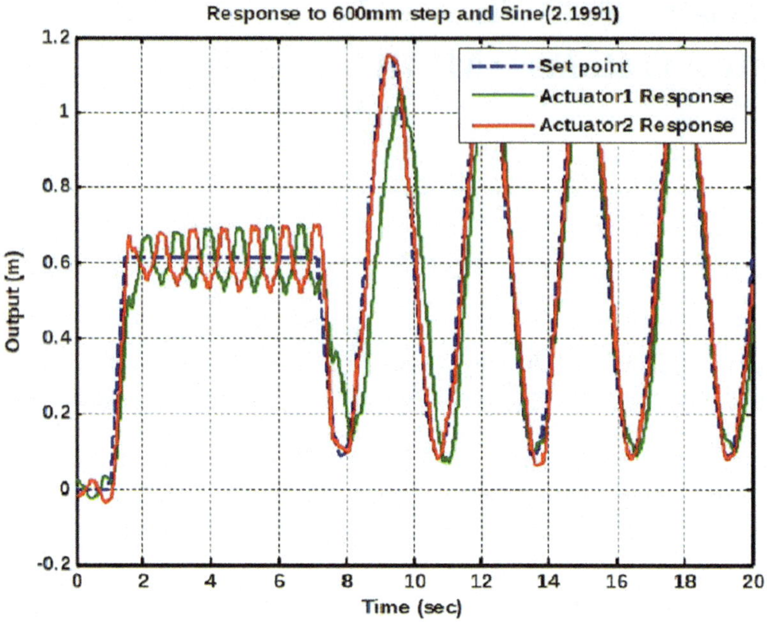

Fig. 7.2 Coupled movement of 2 DOF [1]

Decoupling provides for a more controlled movement (Fig. 7.3).

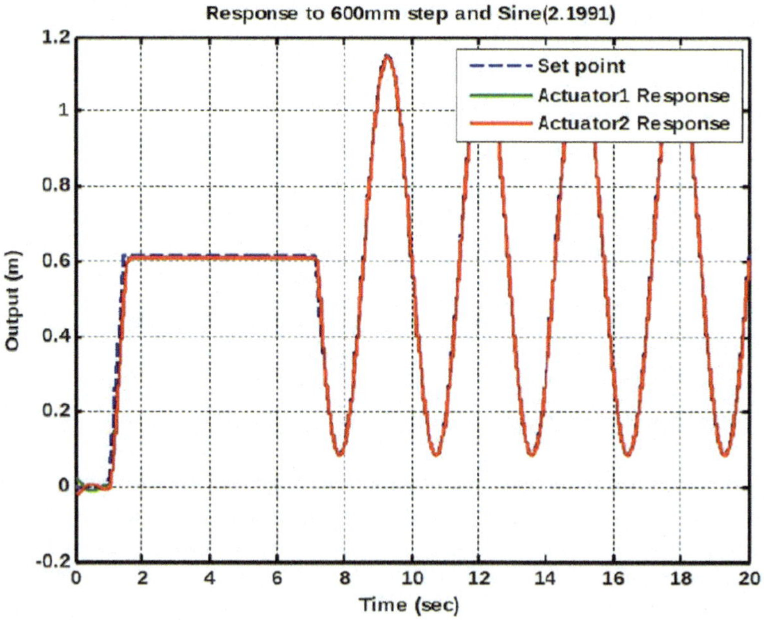

Fig. 7.3 Controlled movement after decoupling [1]

Coupling occurs when multiple actuators drive a common load. Coupling also occurs when the output of a drive affects the input to the controller of another drive. There are degrees of coupling, such as fully coupled, partially coupled, and decoupled systems.

The purpose of decoupling is to facilitate control of MIMO to reduce coupling to a tolerable level. There are metrics for assessing the degree of coupling that are essential for the design of decouplers.

Reference [2] tracks the development of the decoupling strategy in several stages:

- Traditional decoupling of linear systems such as diagonally dominant matrix, relative gain analysis, single value decomposition, and dyadic expansion.
- Modern decoupling for nonlinear systems such as adaptive decoupling, variable structure, robust decoupling, state variable decoupling, and output feedback.
- Intelligent decoupling with neural networks and fuzzy decoupling.
- Advanced intelligent decoupling, which contains adaptive control, predictive control, neural network, fuzzy, genetic algorithm, etc.

A review of all Industrial MIMO decoupling algorithms is given in [3] Much research is still under way on this topic of decoupling control. An attempt is made here to cite a few of the above.

7.2 Back-to-Back Actuator Test Setup

A common setup to test EH servo position control systems for performance against load is a back-to-back arrangement with a force control system (Fig. 7.4).

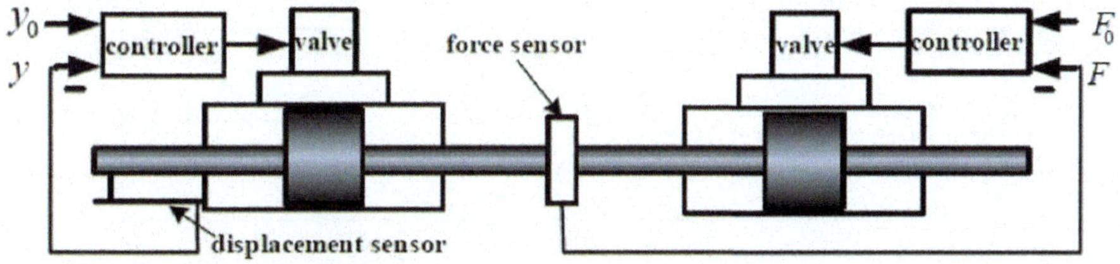

Fig. 7.4 Test setup for decoupling study [4]

The setup presents an ideal study for decoupling [4].

The basic system block diagram, diagonal matrix-based decoupling scheme are shown below (Fig. 7.5).

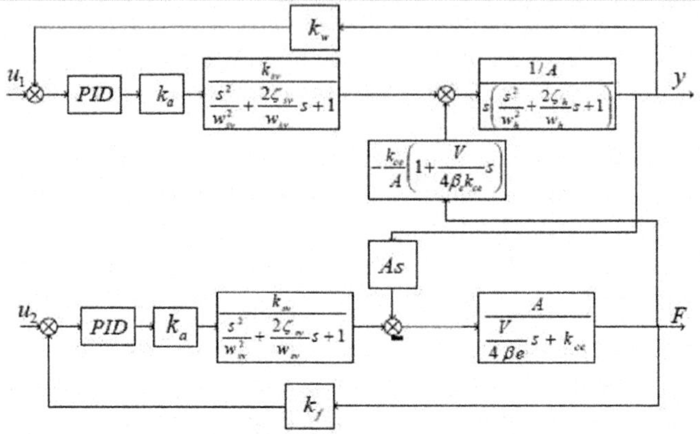

. Block diagram of test system.

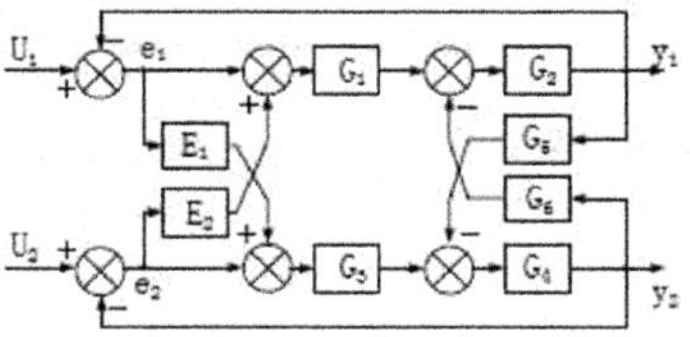

. The schematic diagram of decoupling control.

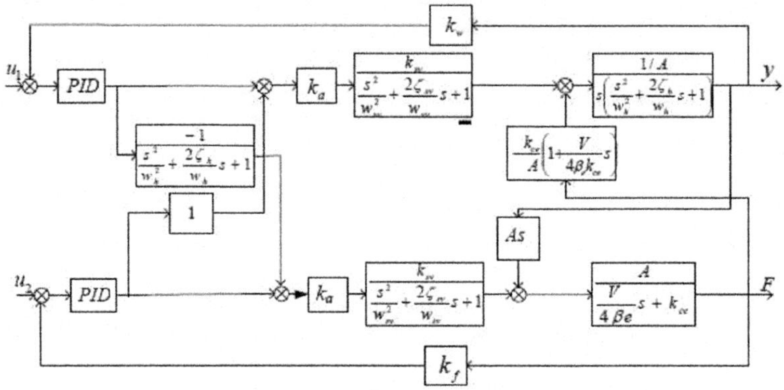

. Transfer function of test system with decoupling link.

Fig. 7.5 Block diagram, decoupling, transfer function schematics [4]

The study was performed using the AMESIM package (Fig. 7.6).

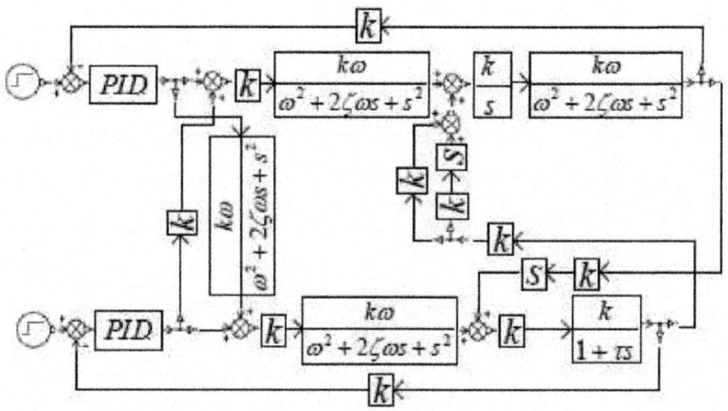

Fig. 7.6 AME simulation model [4]

The effect of a 25 mm step position input on the force set at 7 kN with and without decoupling is shown here (Fig. 7.7).

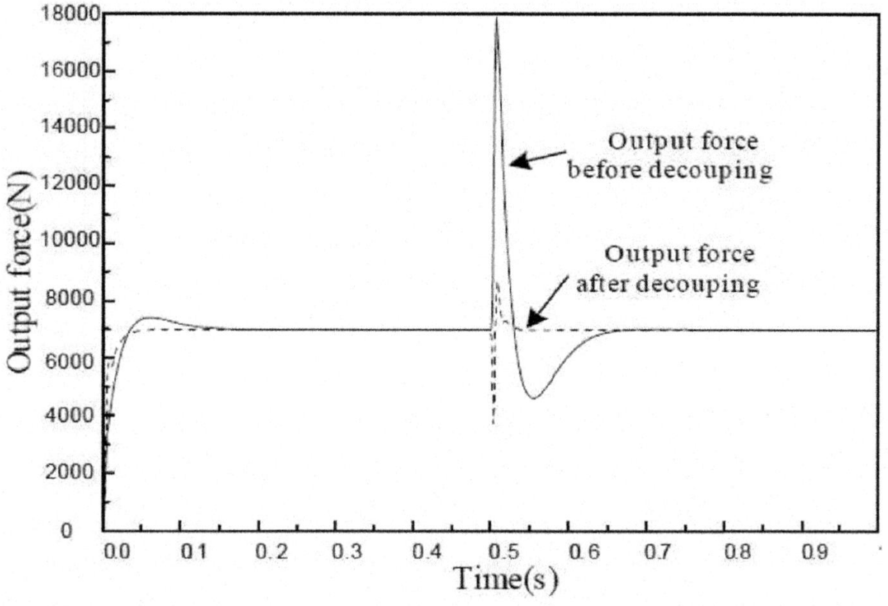

Fig. 7.7 Effect of decoupling [4]

While decoupling drastically reduces force fluctuation, it is still not ideal, as the force loop response is faster than the position response.

A lag/lead network is introduced in the force loop to reduce its response (Figs. 7.8, 7.9, and 7.10).

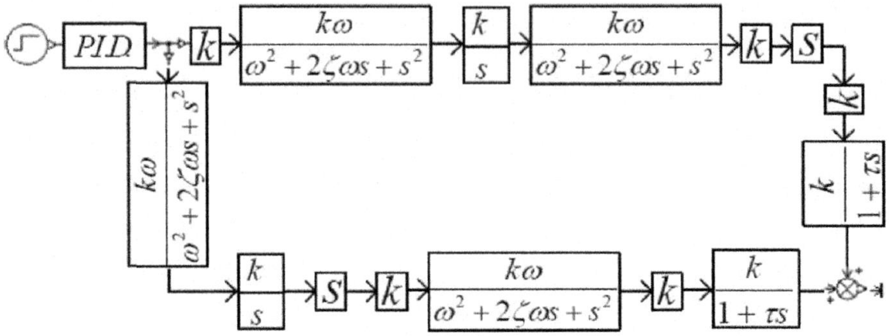

Fig. 7.8 AME sim model with lag/lead network in force loop [4]

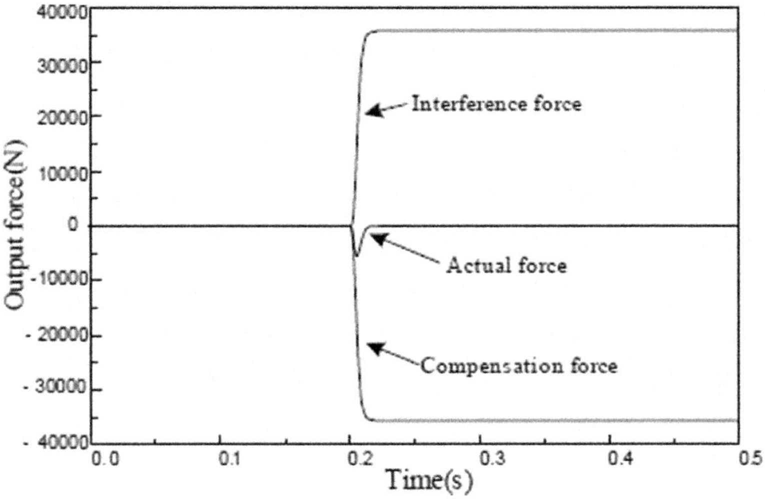

Fig. 7.9 Force variation with step position input [4]

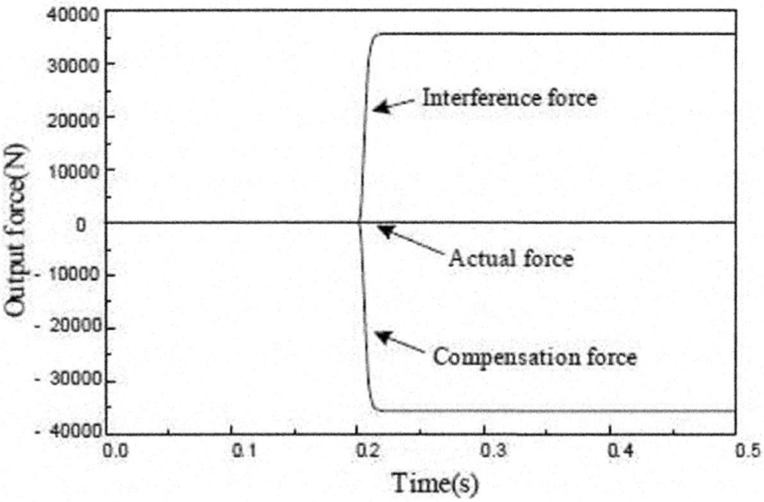

Fig. 7.10 Effect after decoupling [4]

Effect of the network in reducing low frequency gain of the force loop is shown in Figs 7.7 and 7.11.

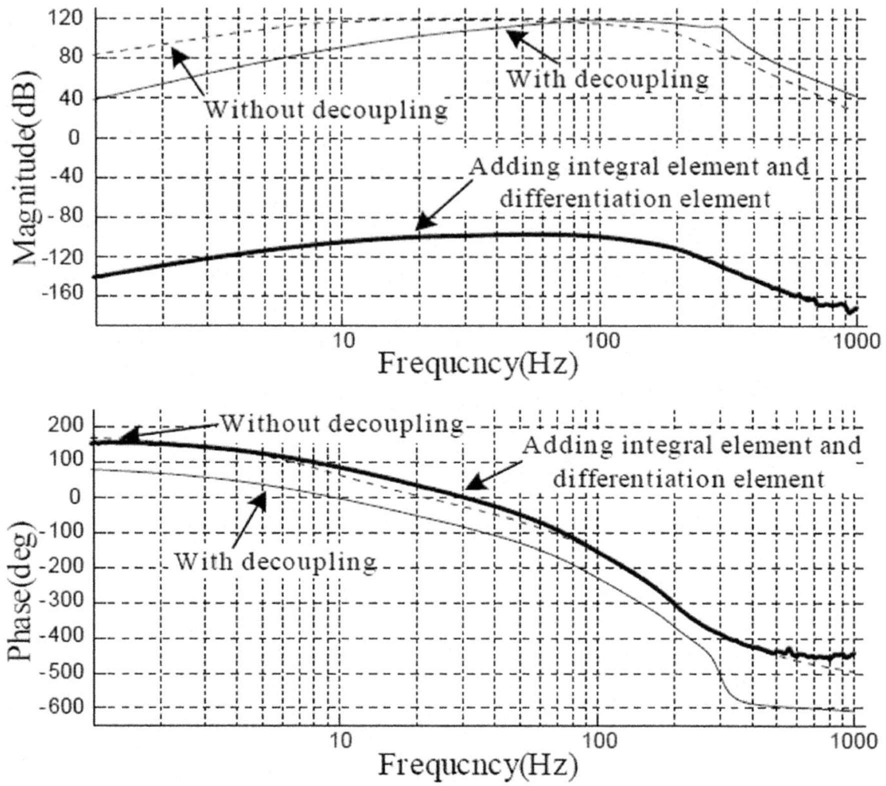

Fig. 7.11 Effect of network in frequency domain [4]

7.3 Experimental Study of 2 DOF Decoupling

Reference [5] presents the results of a study on 3 traditional decoupling strategies—Eigenvalue, single value, and dyadic transfer matrix decoupling methods carried out on a special 2 DOF test stand (Fig. 7.12).

Fig. 7.12 2 DOF test stand [5]

The setup consists of a horizontal actuator driving a guided table and a vertical actuator moving a structure about a pivot connected to a horizontal actuator piston (Figs. 7.13, and 7.14).

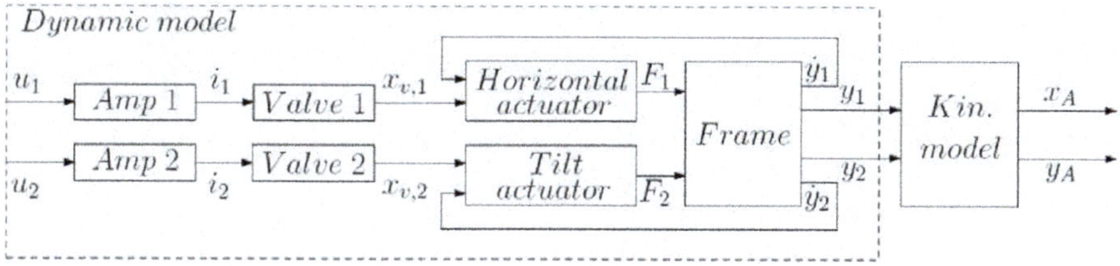

Fig. 7.13 Block diagram of system [5]

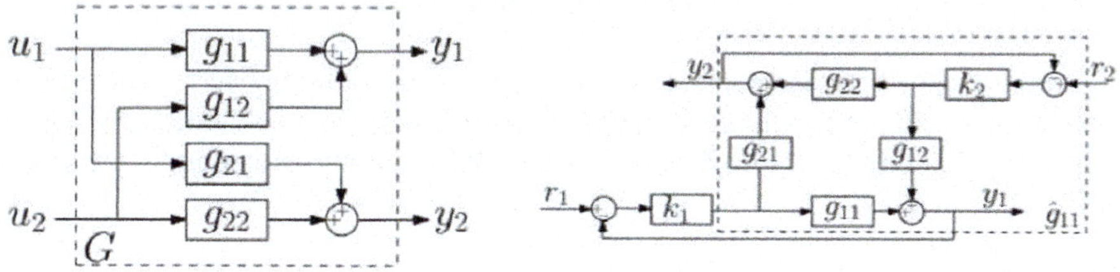

Fig. 7.14 2 DOF block diagram showing coupling between axes [5]

Interactive terms g12 and g21 define the degree of coupling, and when one of them is 0, the system is decoupled. r1 and r2 are the transformed inputs to the system.

After completely identifying the system's kinematic and dynamic parameters, the transfer function matrix was identified by multivariable frequency response measurements at different positions of the end effector. Sensitivity measurements are performed by injecting uncorrelated noise signals $\omega 1$ and $\omega 2$ and measuring plant inputs $u1$ and $u2$ (Fig. 7.15).

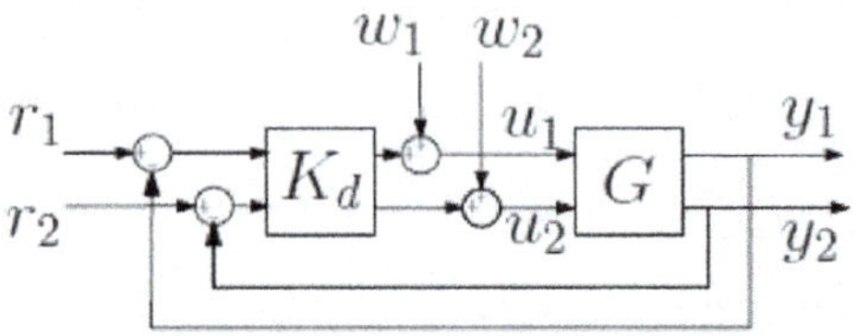

Fig. 7.15 Multivariable frequency response for the transfer function matrix [5]

Considering the variation of hydraulic stiffness with actuator stroke, the FRF matrix is performed at different stroke combinations of actuators.

The interaction coefficients studied indicated maximum coupling effects at approximately 25 Hz.

Eigenvalue decoupling, single value decoupling (SVD), and a dyadic transfer function matrix were implemented, and the results are shown in Fig. 3.5 (Fig. 7.16).

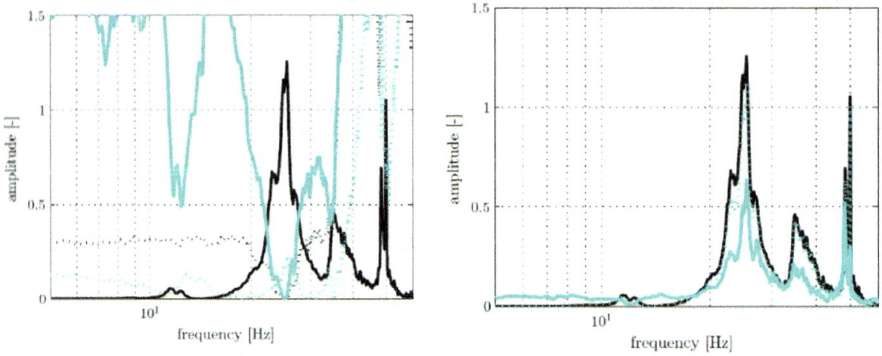

Fig. 7.16 Results of the decoupling study in one configuration [5]

A comparison between manual (left) and optimal decoupling (right). The original interaction coefficient (bold solid) and interaction coefficients obtained after eigenvalue (bold dotted), SVD (gray solid), and DTM decoupling (gray dotted) are shown in Fig. 5.

The authors conclude that SVD shows effective decoupling.

7.4 Hot Strip Rolling Mill

Reference [6] addresses the coupling that exists between the ASC and the tension controller in the hot strip mill process (Figs. 7.17, and 7.18).

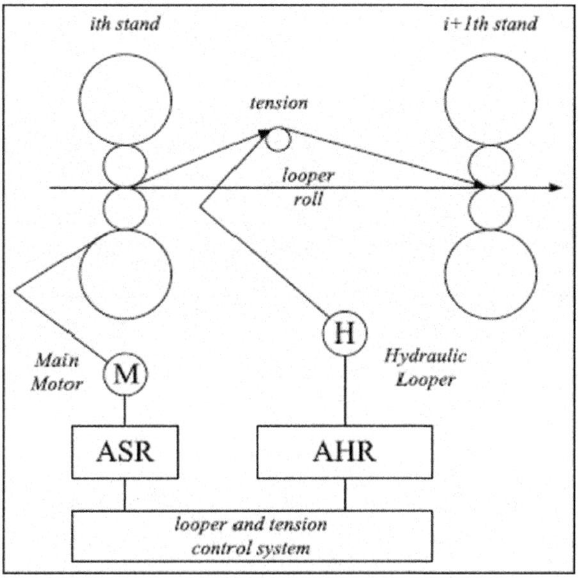

Fig. 7.17 Hot strip mill process [6]

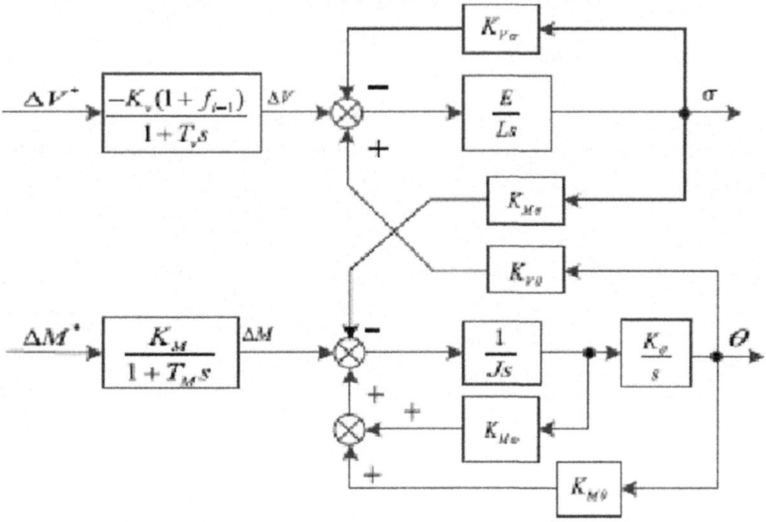

Fig. 7.18 Coupled model of a looper/tension system [6]

The design of the decoupler is founded on the weighting functions as in Fig. 3 (Fig. 7.19).

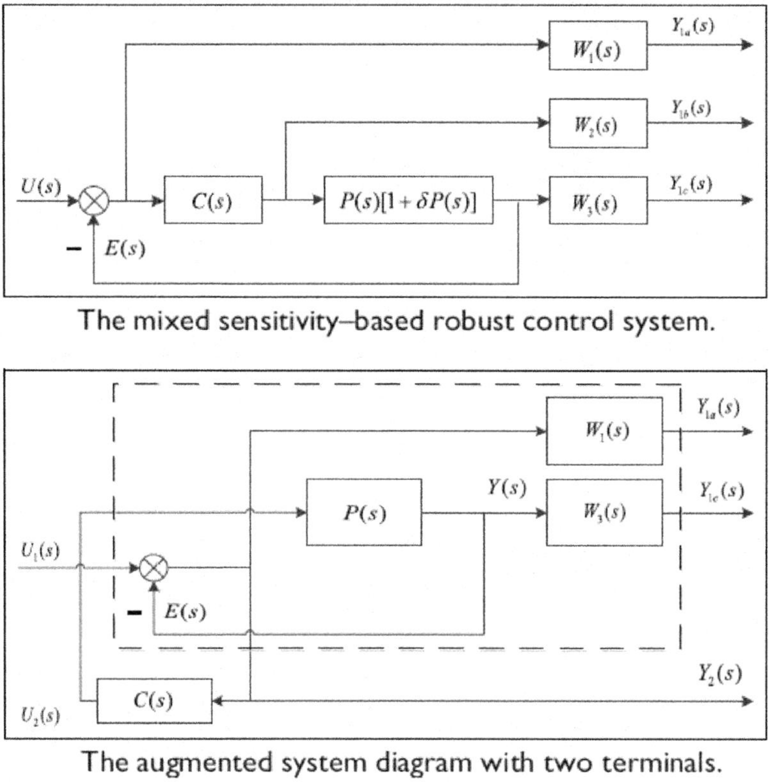

Fig. 7.19 Robust control system [6]

The authors propose a selection method for the weighting function to realize a robust decoupling performance for controller implementation simultaneously (Fig. 7.20).

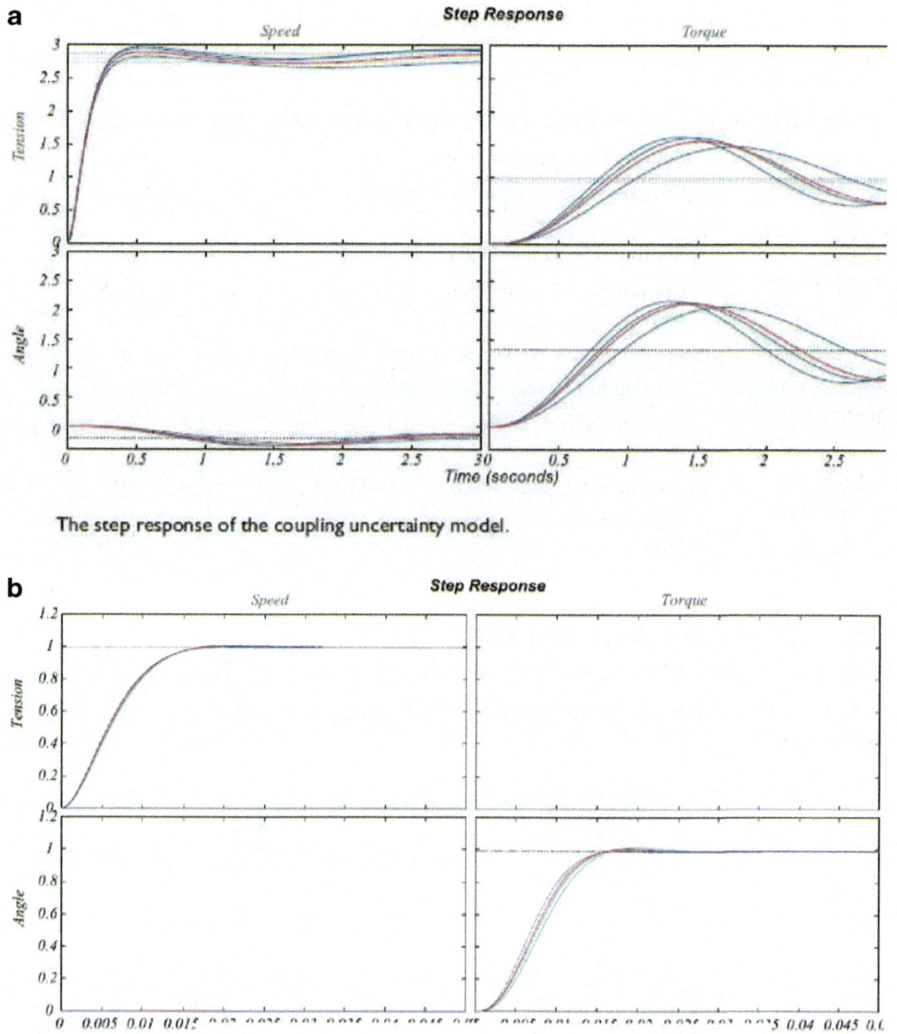

Fig. 7.20 **a** Step response of the coupling uncertainty model [6], **b** response with robust decoupling controller [6]

The performance of the robust controller is recorded in Table 7.1.

Table. 7.1 Performance of the robust controller with decoupler [6]

Control performance		Proposed approach	Current approach
The control performance comparison with 3.5-mm thickness and 10.5-m/s rolling speed			
Case 1: 1.0 MPa perturbation	Settling time (tension loop)	≤50 ms	≤48 ms
	Coupling effect (angle loop)	0.05°	0.5°
Case 2: 2 0 perturbation	Settling time (angle loop)	≤60 ms	≤57 ms
	Coupling effect (tension loop)	Almost zero	0.1 MPa
The control performance comparison with 2.85-mm thickness and 1 3-m/s rolling speed			
Case 1: 1.0 MPa perturbation	Settling time (tension loop)	≤55 ms	≤62 ms
	Coupling effect (angle loop)	0.2°	2°
Case 2: 2.0' perturbation	Settling time (angle loop)	≤66 MPa	≤75 MPa
	Coupling effect (tension loop)	0.05 MPa	0.15 MPa

7.5 Automotive Axle Test Rig

Reference [7] addresses the decoupling of an automotive test rig for durability tests on an axle (Fig. 7.21).

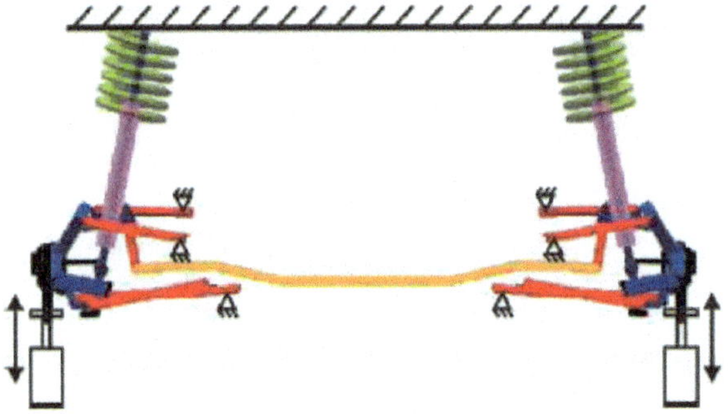

Fig. 7.21 Axle test rig for durability [7]

The study proposes a decoupling procedure to render the relation between transformed inputs and outputs as diagonal as possible.

Then, decentralized controller design techniques are used for individual SISO controllers.

The study proposes a decoupling procedure to render the relation between the transformed inputs and outputs as diagonal as possible. Then, decentralized controller design techniques are used for the individual SISO controllers (Figs. 7.22, 7.23 and 7.24).

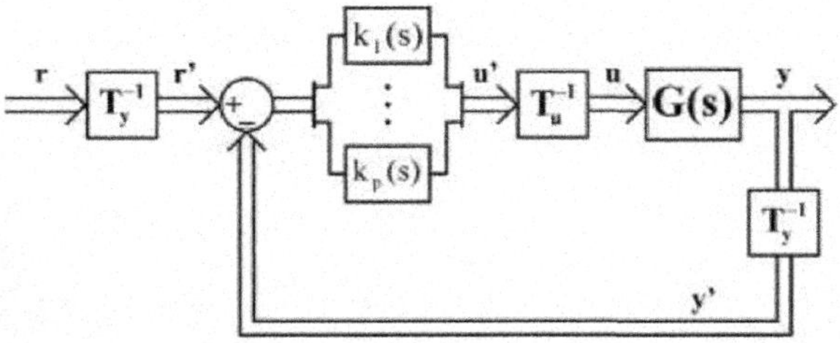

Fig. 7.22 Decoupling transformation of inputs and outputs and p-independent SISO controllers [7]

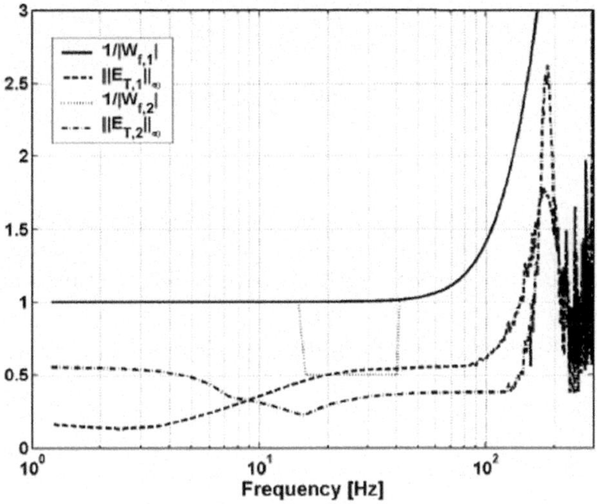

Decoupling results: $|W_{f,1}|$ used in first decoupling optimization (full line) and resulting difference between $\tilde{\mathbf{G}}_\mathbf{d}$ and $\tilde{\mathbf{D}}_\mathbf{d}$: $\|\mathbf{E}_{\mathbf{T},\mathbf{1}}\|_\infty$ (dashed line). $|W_{f,2}|$ (dotted line) and $\|\mathbf{E}_{\mathbf{T},\mathbf{2}}\|_\infty$ (dash-dotted line) are the same for the second decoupling optimization.

Fig. 7.23 Results of decoupling optimization [7]

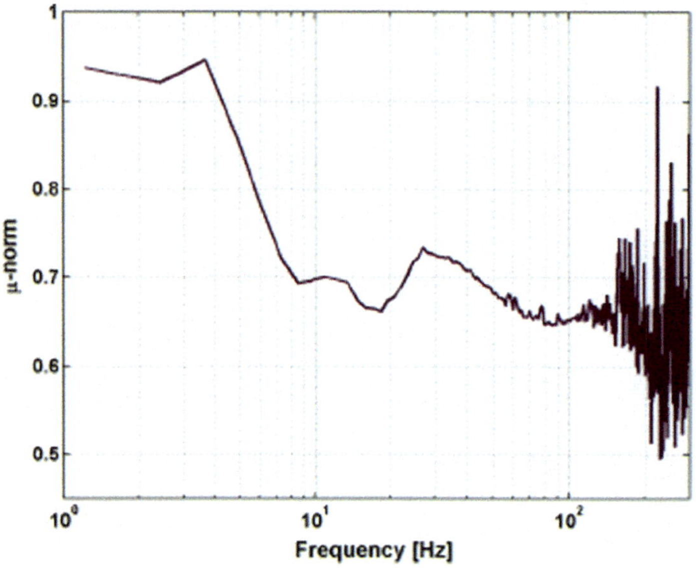

Fig. 7.24 Robustness analysis $\mu < 1$ at all frequencies [7]

This calculation is aided by the measurement of the frequency response matrix FRF matrix, noting no identification of the MIMO model. The decentralized robust design is performed as for a SISO system. Owens dyadic transformation matrices are used to calculate initial estimates, and a frequency-dependent weighting function is proposed for optimization of decoupling.

7.6 Tracked Vehicle

Tracked vehicles driven by individual hydrostatic drives enable higher forward speeds when motors run at the same speeds and sharp steering for maneuverability when motors run at different speeds (Fig. 7.25).

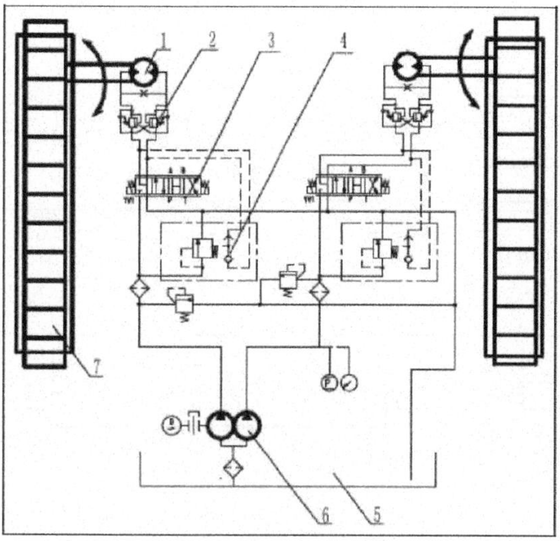

1. Hydraulic motor
2. Balance valve,
3. Servo valve,
4. Three-way compensator,
5. Hydraulic oil tank,
6. Double pump,
7. Track

Fig. 7.25 System under study [8]

The configuration demands decoupling to meet these objectives. Reference [8] applies QFT for robust control of the vehicle and investigates control effectiveness in steering control of the vehicle (Figs. 7.26, 7.27 and 7.28).

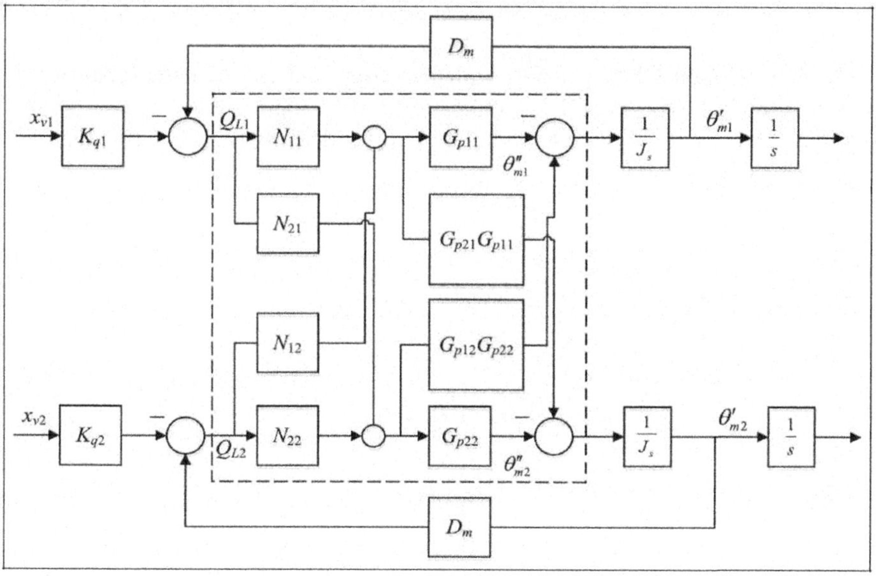

Fig. 7.26 Block diagram of drive system decoupling [8]

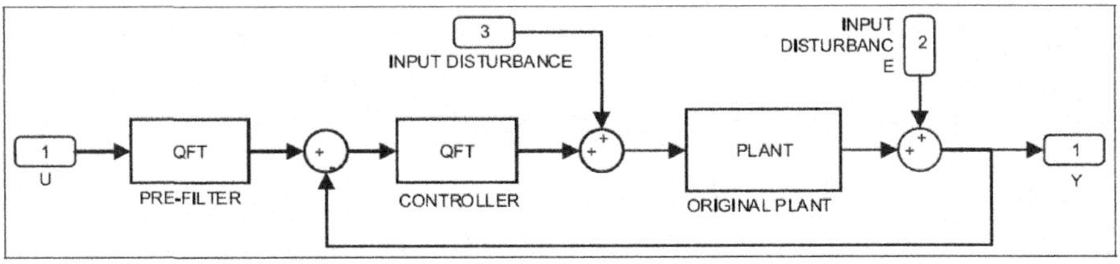

Fig. 7.27 Structure of robust QFT control [8]

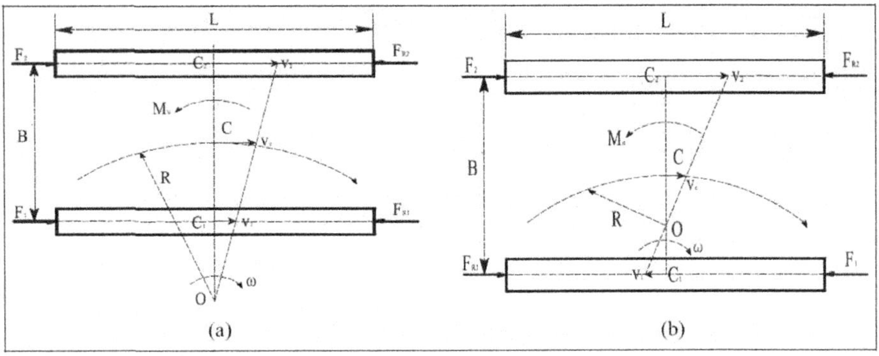

Fig. 7.28 Steering types. Differential and in-situ steering [8]

The design of the QFT controller with decoupling and transient disturbance rejection is combined with the QFT prefilter to provide robustness. The two states of steering control were verified by simulation to validate the QFT control (Figs. 7.29 and 7.30; Table 7.2).

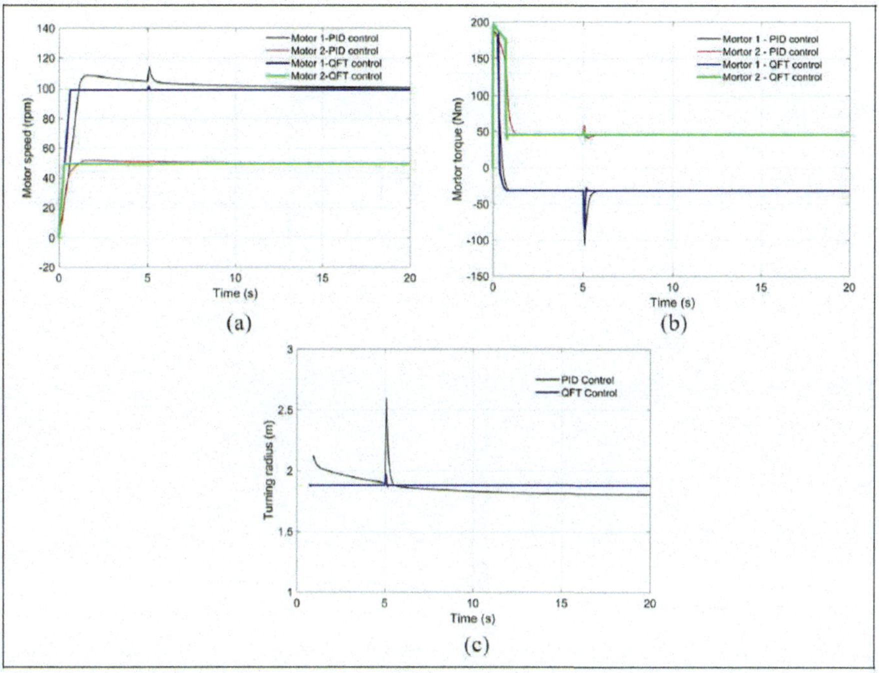

$R \geqslant B/2$ steering states.

Fig. 7.29 Simulation output [8]

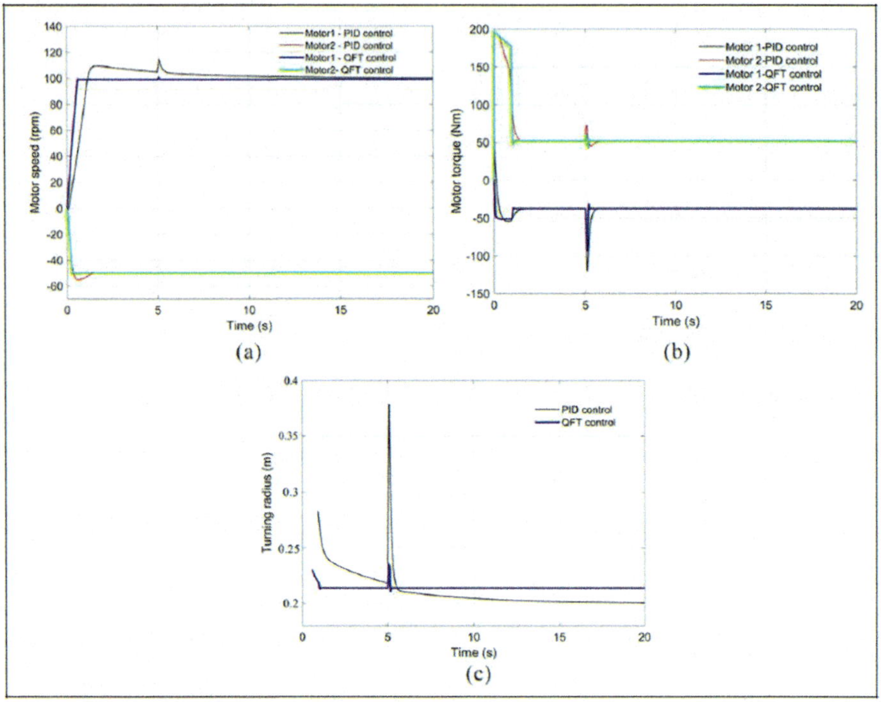

$0 \leq R < B/2$ steering states.

Fig. 7.30 Simulation result [8]

Table. 7.2 Parameters of vehicle for simulation [8]

Parameter	Numerical value
Tracked vehicle mass (m/kg)	500
Moment of inertia of vehicle (J/kg m^2)	550
Body width (B/m)	1.2
Track length (L/m)	1.22
Radius of motor drive wheel (r/m)	0.095
Rolling friction coefficient, f	0.05
Steering drag coefficient, μ	0.6
Overall efficiency of hydraulic motor and transmission mechanism η	0.63
Side transmission ratio, i	1
Servo valve flow gain (K_q/s V)	0.013
Servo valve flow pressure coefficient	2.06271×10^{-10}
K_c/m^3/s Pa	2.5×10^{-12}
Hydraulic motor leakage coefficient, C_t	
Oil volume elastic modulus, β_e	1000
Hydraulic motor inertia (J_i/kg/m^3)	10
Oil density (ρ/kg/m^3)	850

The steering performance was verified experimentally with high-precision motion track instrumentation (Figs. 7.31 and 7.32; Table 7.3).

Fig. 7.31 Test site [8]

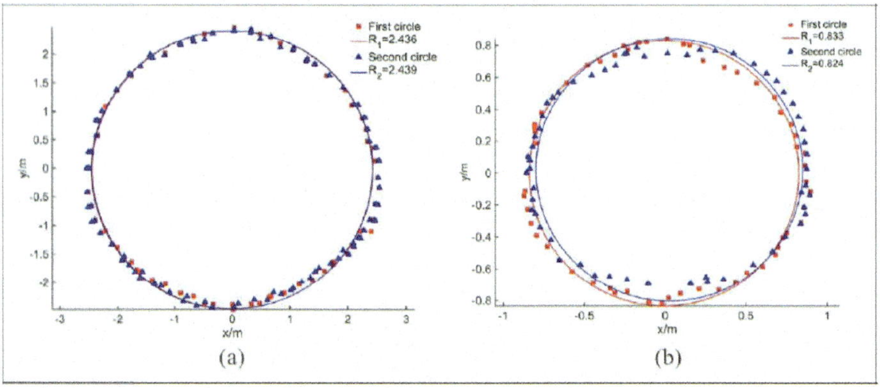

Turning tracks of the tracked vehicle: (a) $R \geqslant B/2$ and (b) $0 \leqslant R < B/2$.

Fig. 7.32 Steering test results [8]

Table. 7.3 Motor speeds during steering tests [8]

Parameter	Differential steering		In-situ steering	
	Inner motor	Outer motor	Inner motor	Outer motor
Theoretical value	50	100	50	−50
Mean	49.83	99.90	49.55	−49.92
Standard deviation	0.40	0.34	0.42	0.34
Coefficient of variation (%)	0.80	0.33	0.80	0.68

7.7 MIMO Controls

MIMO controls are necessary to meet the stability and tracking accuracy of systems moving in free space or in constrained space. The metric for the comparison of various methods is emax/max speed.

MIMO controls are broadly classified as centralized and decentralized controllers. A few of these controls are cited herein.

7.7.1 Free Space

7.7.1.1 VDC, SMISMO

The virtual decomposition control VDC approach proposed by Zhu uses subsystem (links/joints) dynamics to conduct control design while guaranteeing stability and convergence. In VDC, the original system is decomposed into subsystems, and the individual systems are designed as in SISO without penalty of system stability. A study of VDC applied to a 3 DOF is detailed in [9].

7.7.1.2 SMC of Crane

MIMO control of a forestry crane working in cartesian space with commands from joysticks is detailed in [10]. The centralized design follows with modeling, control, simulation, and testing steps.

The overall kinematic design is based on the cartesian workspace and desired speeds. The dynamics causing joint torques by hydraulic actuator forces and geometry result in load handling (Fig. 7.33).

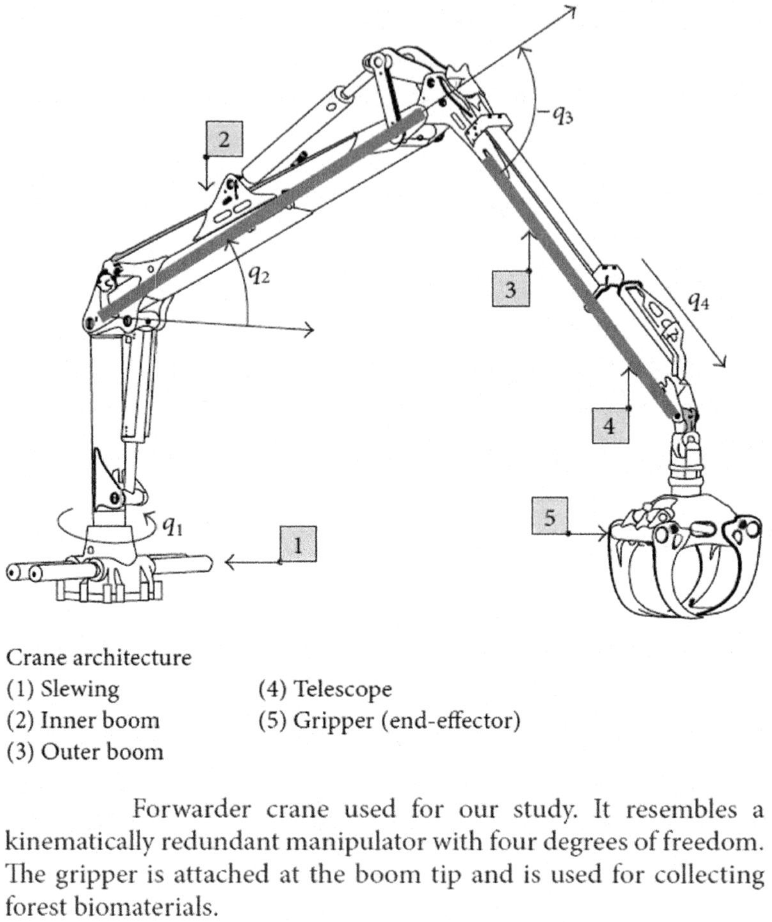

Fig. 7.33 Forestry crane [10]

Hydraulic force includes friction forces as well as drive dynamics. The model is corrected by friction measurements. To cope with model uncertainties and disturbances, two variations of robust sliding mode controllers were designed using the setup in Fig. 7.34.

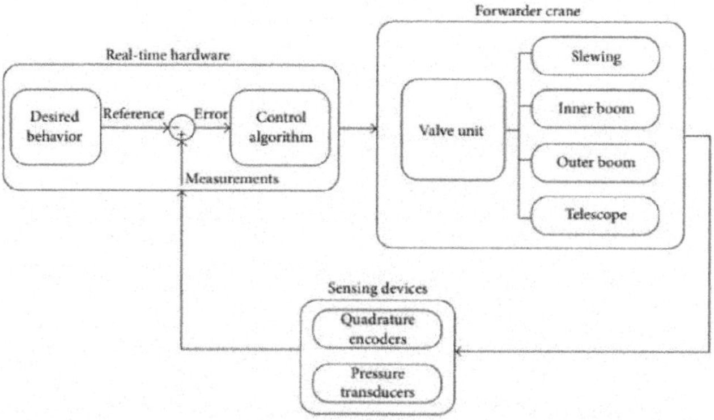

The layout presents the configuration of our control system setup, which includes the measurement devices and the rapid prototyping unit. The valve is equipped with magnetic coils to provoke the motion of the internal spools and a power amplifier that accepts regular PWM signals.

Fig. 7.34 Setup [10]

The setup provides 3 angular encoders, 1 linear transducer, and 2 pressure transducers for each of the actuators. Parameter estimations from measurements are performed using the Kalman filter (Figs. 7.35 and 7.36).

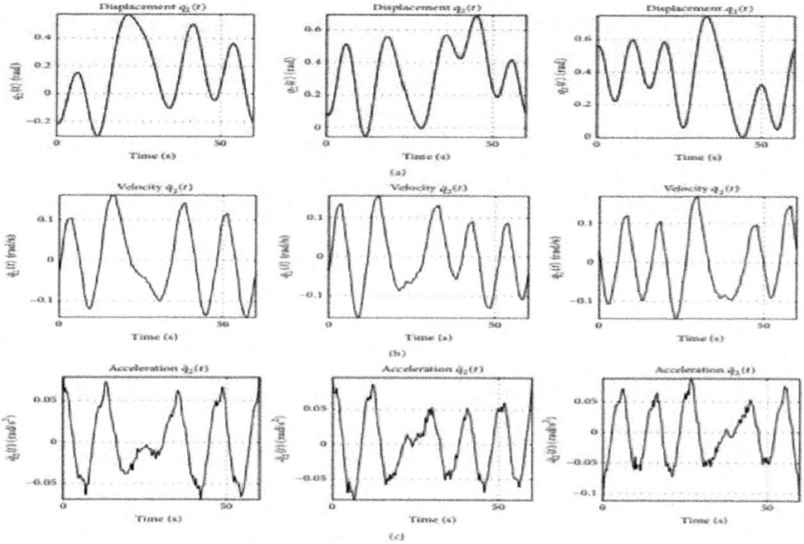

Example of recorded trajectories which resemble finite Fourier series. Each column represents a different data set, which results from an average of many periods. (a) Measured angular position $q_2(t)$ in (rad). (b) Estimated angular velocity $\dot{q}_2(t)$ in (rad/sec). (c) Estimated angular acceleration $\ddot{q}_2(t)$ in (rad/sec^2).

Fig. 7.35 Joint displacement, velocity, acceleration [10]

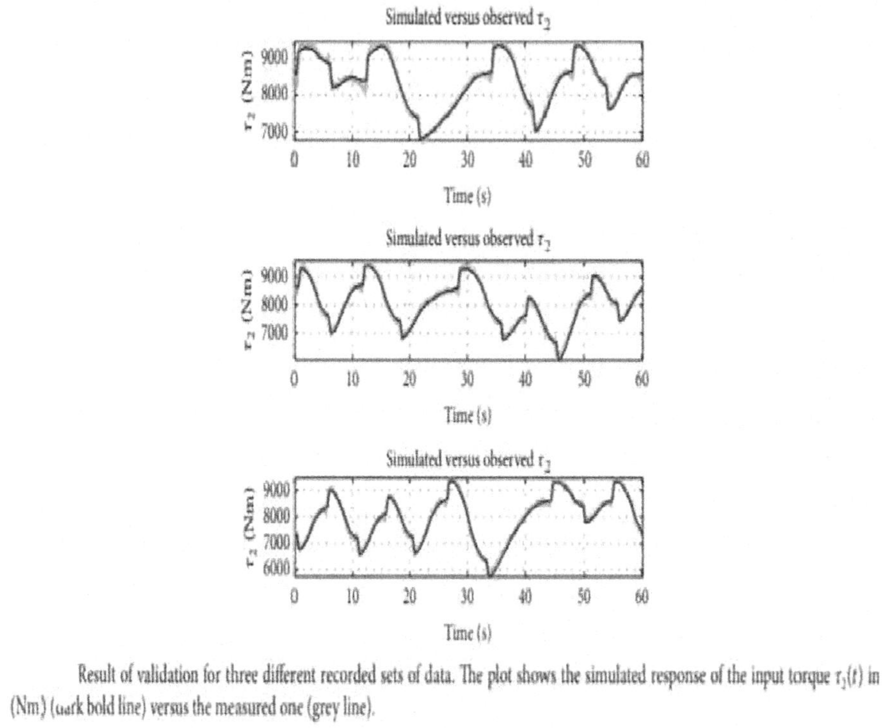

Fig. 7.36 Joint torques—simulated versus measured [10]

The results of tracking for first-order sliding and second-order sliding are shown in Figs. 7.37 and 7.38.

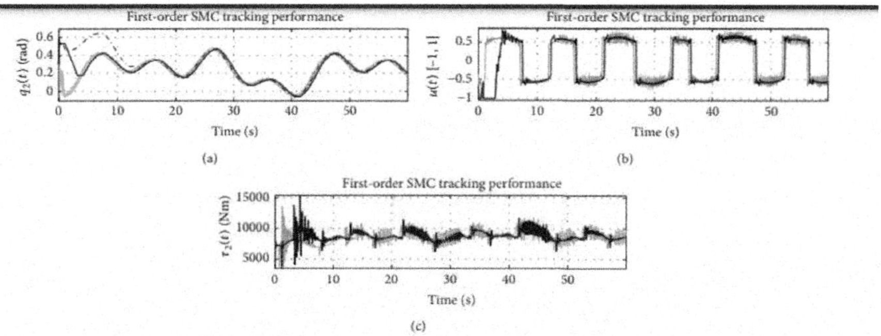

: Experimental results of tracking performance with a first-order SMC. (a) *Angular link displacement in radians.* The bold signal shows the results applying the chattering reduction method (47), the striped signal shows results with the original formula (36), and the gray signal is the reference trajectory. (b) *Control input.* The bold signal shows the results applying the chattering reduction method (47) and the gray signal shows results with the original formula (36). (c) *Computed torque from pressure measurements.* The bold signal shows the results applying the chattering reduction method (47) and the gray signal shows results with the original formula (36).

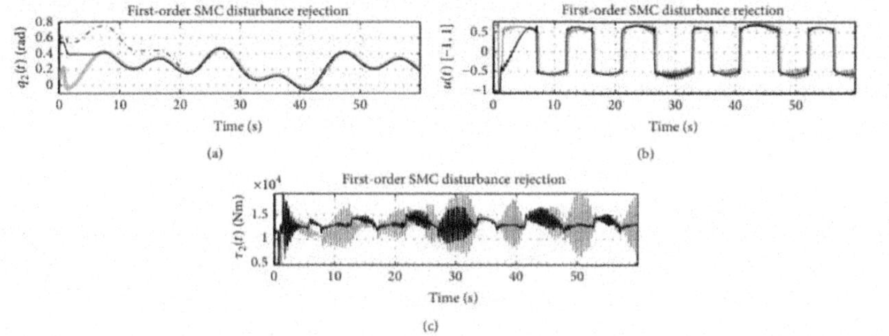

: Experimental results of disturbance rejection with a first-order SMC. (a) *Angular link displacement in radians.* The bold signal shows the results applying the chattering reduction method (47), the striped signal shows results with the original formula (36), and the gray signal is the reference trajectory. (b) *Control input.* The bold signal shows the results applying the chattering reduction method (47) and the gray signal shows results with the original formula (36). (c) *Computed torque from pressure measurements.* The bold signal shows the results applying the chattering reduction method (47) and the gray signal shows results with the original formula (36).

Fig. 7.37 Results of first-order SMC [10]

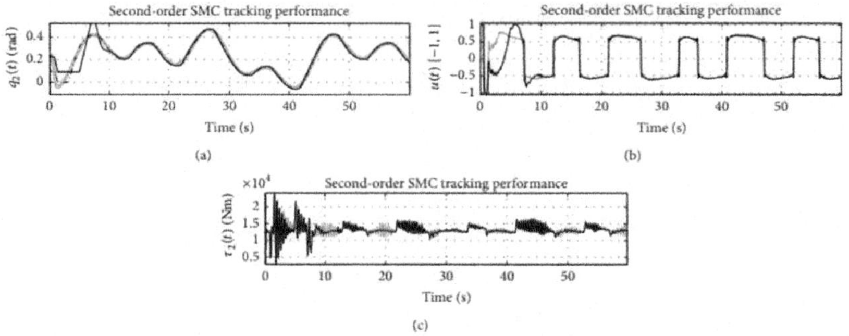

Experimental results of tracking performance with a second-order SMC. (a) *Angular link displacement in radians.* The bold signal shows the results applying the state-dependent method (48), the striped signal shows results with the original formula (40), and the gray signal is the reference trajectory. (b) *Control input.* The bold signal shows the results applying the state-dependent method (48) and the gray signal shows results with the original formula (40). (c) *Computed torque from pressure measurements.* The bold signal shows the results applying the state-dependent method (48) and the gray signal shows results with the original formula (40).

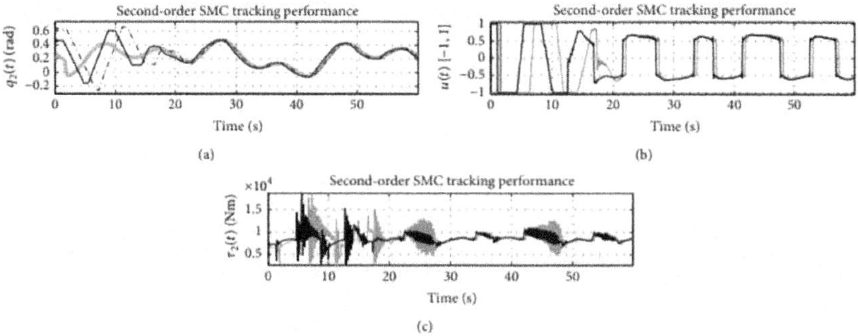

i: Experimental results of disturbance rejection with a second-order SMC. (a) *Angular link displacement in radians.* The bold signal shows the results applying the state-dependent method (48), the striped signal shows results with the original formula (40), and the gray signal is the reference trajectory. (b) *Control input.* The bold signal shows the results applying the state-dependent method (48) and the gray signal shows results with the original formula (40). (c) *Computed torque from pressure measurements.* The bold signal shows the results applying the state-dependent method (48) and the gray signal shows results with the original formula (40).

Fig. 7.38 Results of second-order SMC [10]

7.7.1.3 Control HyQ Leg

Reference [11] details the control of a 2-legged (hip, knee) hydraulic robot used in the development of a quadruped leg (Figs. 7.39, 7.40 and 7.41).

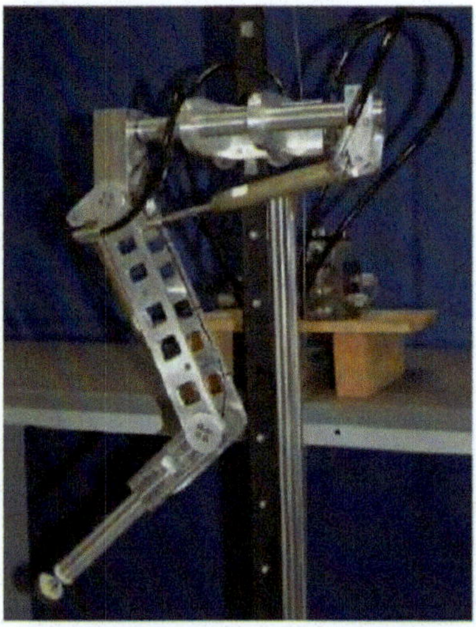

Fig. 7.39 Hip/knee-jointed rig on upright stand [11]

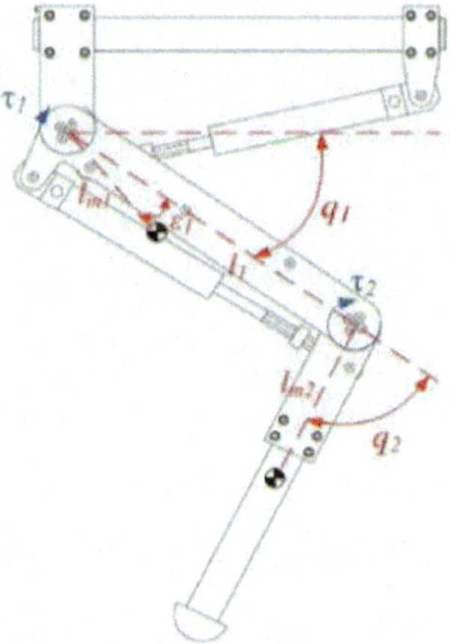

Fig. 7.40 Geometry of bionic leg

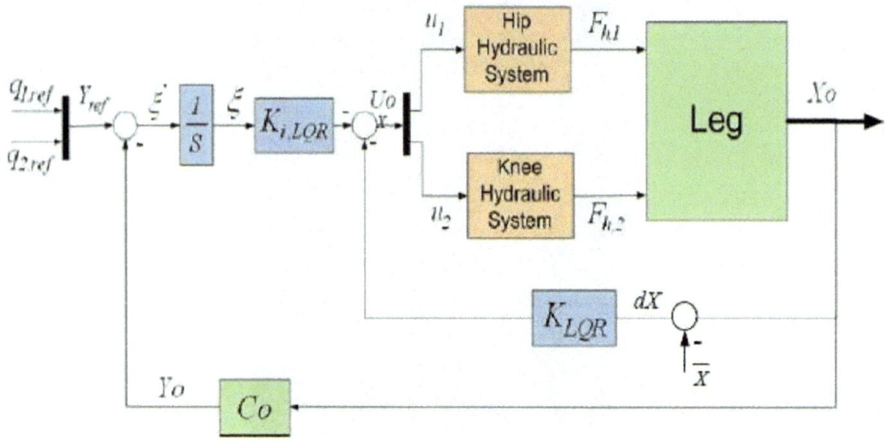

Fig. 7.41 Block diagram of the LQR controller with integrators [11]

A control block diagram of the system is shown in Fig. 3.

PID, LQR and feedback linearization FL, control methods were applied, and the response to a 1 Hz sinusoidal input is compared (Figs. 7.42 and 7.43).

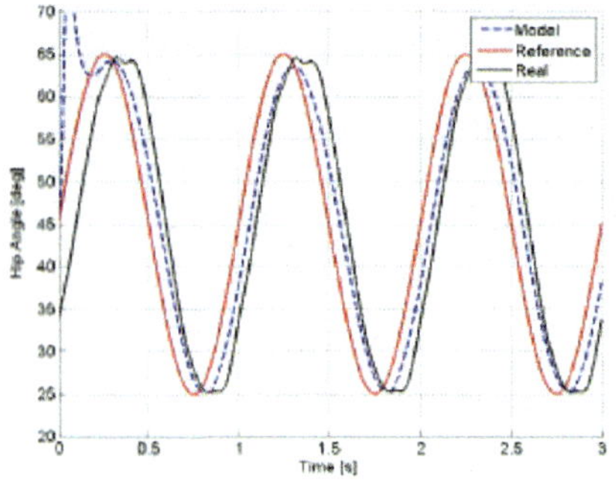

. PID sinusoidal response – model (blue) – real (black).

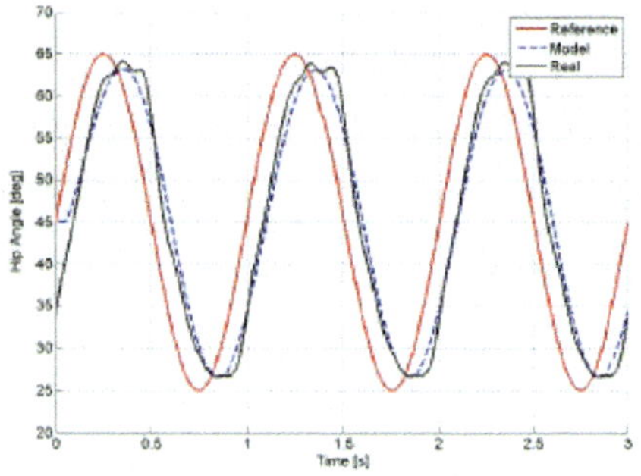

. LQR sinusoidal response – model (blue) – real (black).

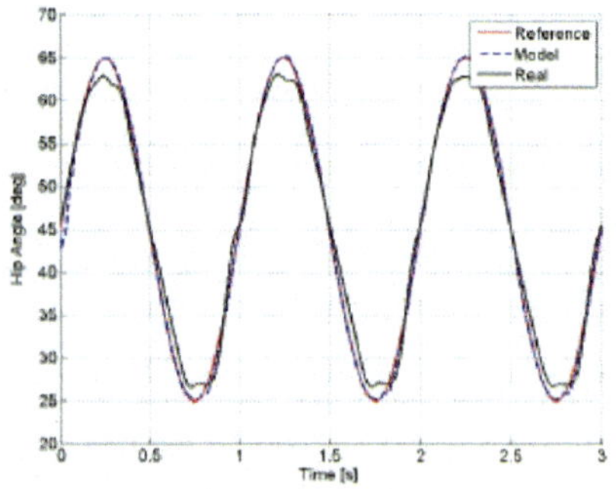

. FL sinusoidal response – model (blue) – real (black).

Fig. 7.42 Response to 1 Hz sine input [11]

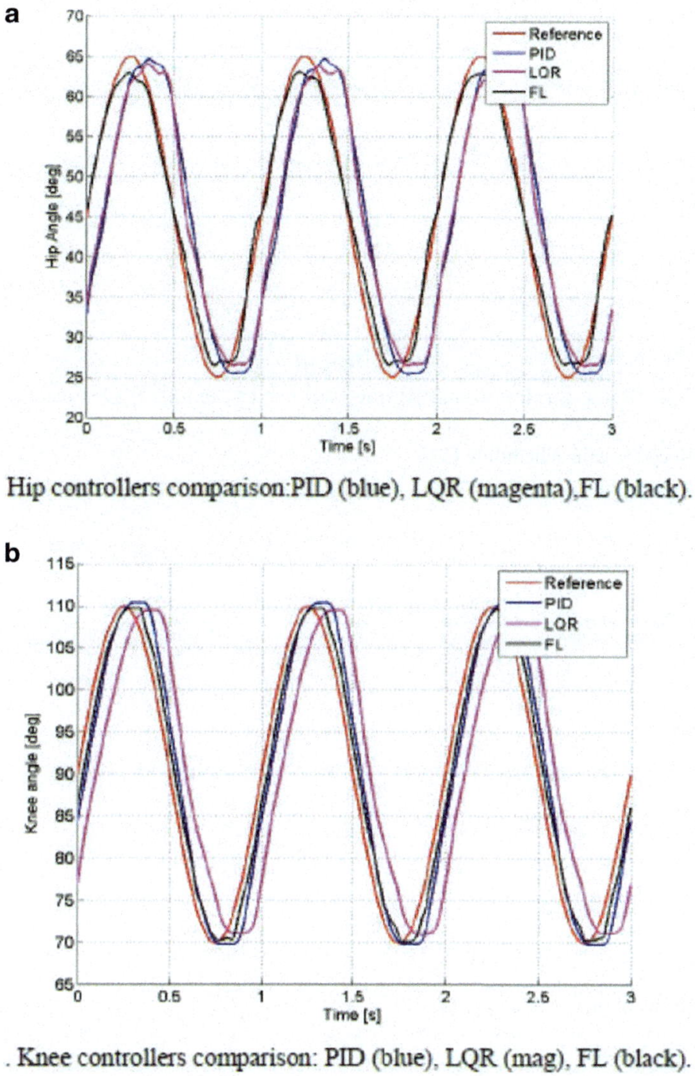

Fig. 7.43 **a** Response of the hip joint [11], **b** response of the knee joint

Reference [12]. The need for gain scheduling on this bionic leg is reported, and the effect of gain scheduling on the response of the leg is compared with that of PID control. (gain scheduling is one specific form of a self-tuning system) (Figs. 7.44 and 7.45).

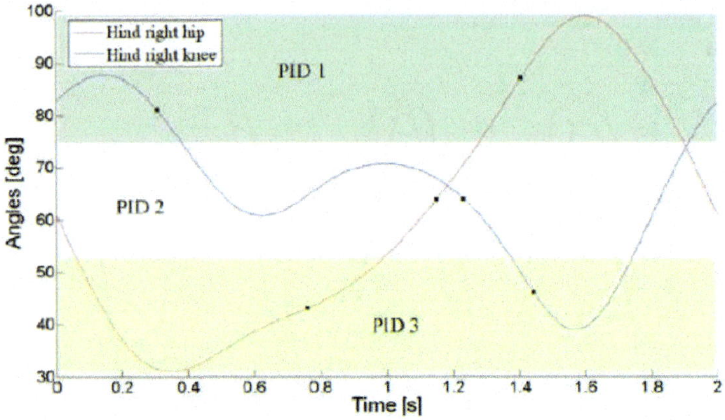

Hip and knee angles in an animal-walking cycle and PID workspaces

Fig. 7.44 Need for gain scheduling [12]

	PID		PID with gain scheduling	
	Hip	Knee	Hip	Knee
IE	0.91	-0.38	-0.29	-0.34
IAE	17.69	9.58	16.77	9.83
ISE	193.07	63.80	181.18	67.15

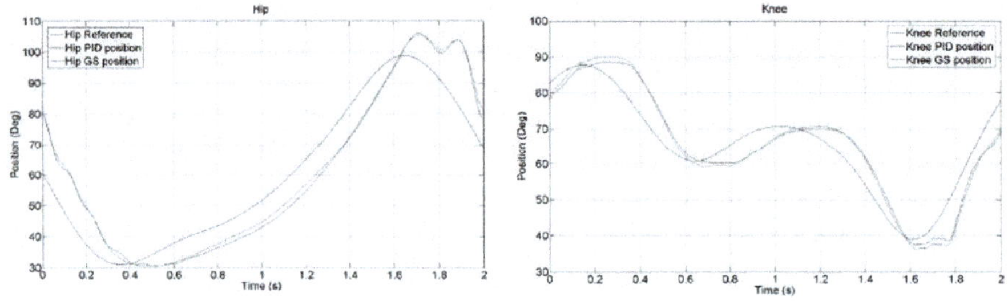

Comparison of trajectory following performance of classic PID vs. PID with gain scheduling (GS)

Fig. 7.45 Effect of gain scheduling [12]

7.7.1.4 Proportional Integral Sliding Mode Control (PI SMC)

SMC is a robust approach to the design of joint drives of a robot manipulator subjected to varying inertial loads and friction forces. To overcome the chattering associated with this control, PI SMC studies for several evolutionary algorithms are presented in [13].

7.7.1.5 Comparison of Control Methods

A survey of tracking performance in free space of different control schemes for robots is presented in [14]. Tracking performance in terms of the ratio of emax to velocity max is the metric.

7.7.2 *MIMO Controls for Tracking in Constrained Space*

7.7.2.1 Force/Impedance Control

Robotic applications for constrained motion, such as deburring and fettling, largely depend on passive compliance at the end effector. Soft mechanical or pneumatic springs with settable preload are used.

The active force or impedance control is obtained by a 6-axis force/moment transducer mounted at the end effector. The control architecture for hybrid control is shown in Fig 7.46 [15].

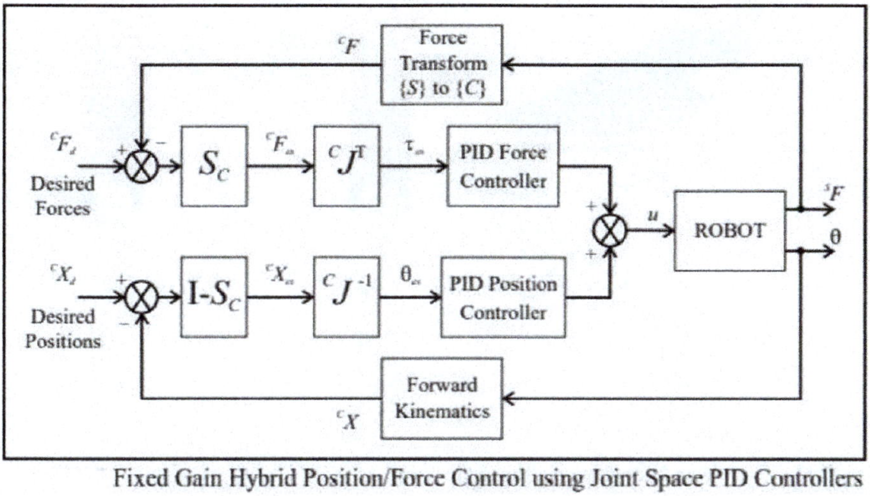

Fig. 7.46 Hybrid control architecture [15]

The output vector of this transducer consists of the force/moment in constrained directions but the position/orientation in orthogonal directions. These cartesian space control schemes are advantageous since the manipulator tasks naturally decompose into this frame of reference, both in terms of desired outputs and control performance criteria [15]

7.7.2.2 Quadruped Robot with Passive Compliance

A quadruped heavy-duty hydraulic robot is designed with passive compliance for its hybrid control of position and force [16]. The design deploys a total of 16 EHS actuators. A schematic of this compliance control is shown in Figs. 7.47 and 7.48.

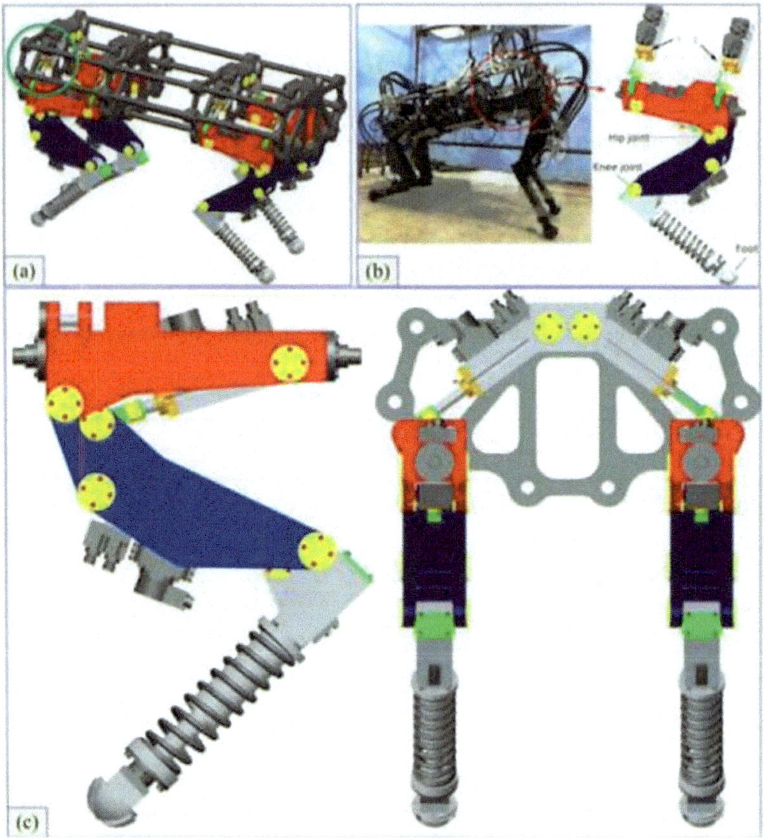

Fig. 7.47 Heavy-duty hydraulic quadruped [16]

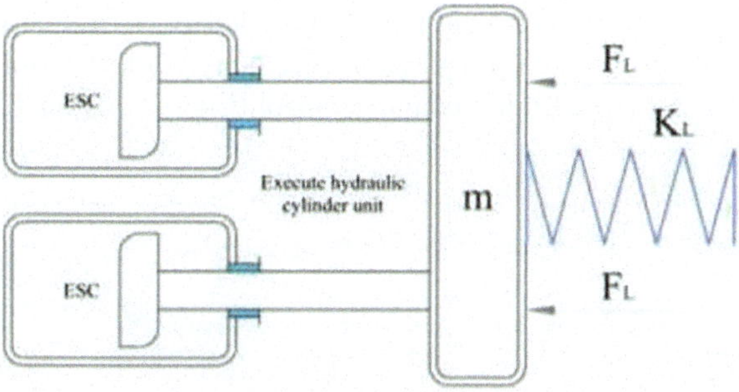

Fig. 7.48 Schematic of compliance

7.7.2.3 Self-tuning PID Control

Unmanned undersea manipulators have weld verification and docking tasks, which call for hybrid position force control. A study of a self-tuning PID control of a TA9 Slings by manipulator is in [15]

7.7.2.4 Impedance Control of HyQ Leg

The HyQ quadruped manipulator is entrusted with highly dynamic tasks needing hybrid position/force control [17] (Fig. 7.49).

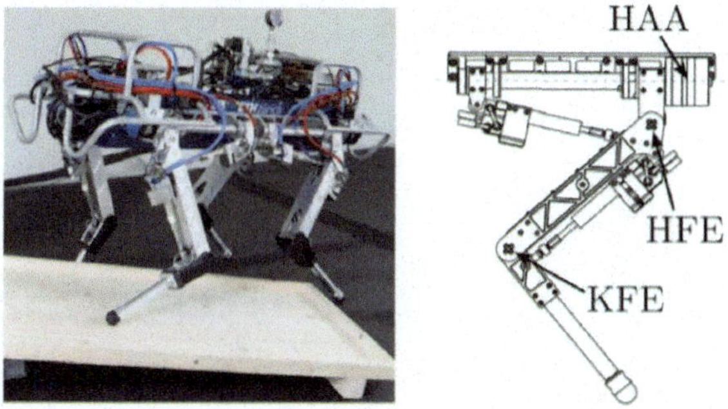

(a) HyQ robot (b) HyQ joint names

(a) The HyQ robot. The HyQ leg drawing in (b) defines the joint names: the hip abduction/adduction (HAA), the hip flexion/extension (HFE), and the knee flexion/extension (KFE).

Fig. 7.49 Impedance control of HyQ leg [17]

Reference [17] describes the design of the impedance controller in two steps: first, the design of the inner force loop with natural velocity compensation and the design of the outer position loop with impedance control (Figs. 7.50 and 7.51).

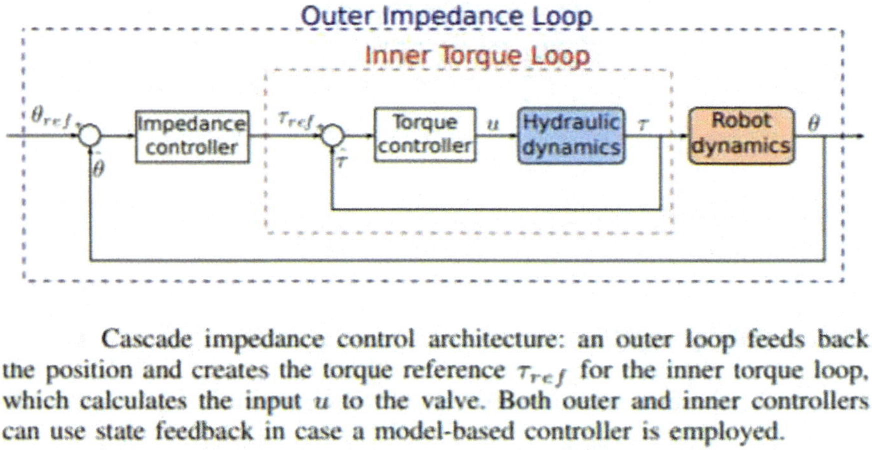

Fig. 7.50 Impedance control scheme [17]

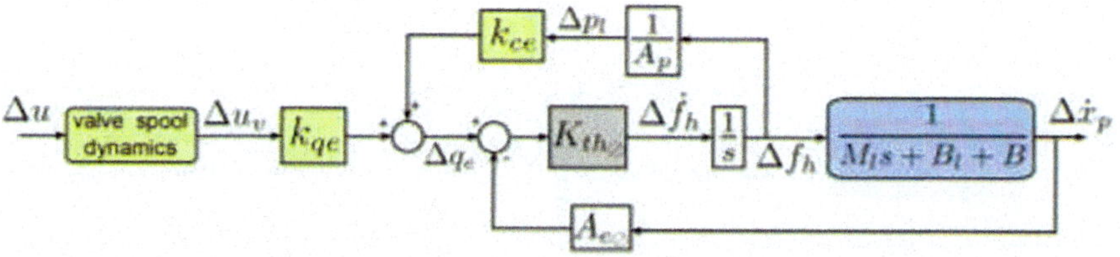

Block diagram for the open-loop *linearised hydraulic force*. The blocks related to the valve are depicted in yellow, the load in blue, and the transmission stiffness in gray. Also, $p_l = p_a - \alpha p_b$ represents the load pressure, and u the valve command. The piston velocity $\Delta \dot{x}_p$ is multiplied by the equivalent area $A_{e\oslash}$ and is transformed into a flow. This flow is then fed back into the force dynamics, representing an intrinsic velocity feedback.

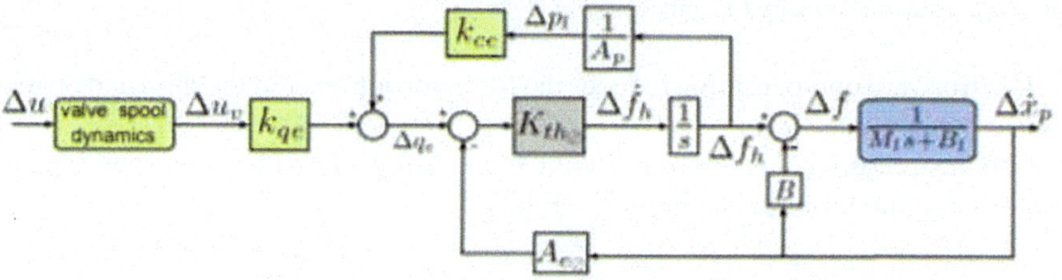

Block diagram for the open-loop *load force* dynamics. Differently from the hydraulic force shown for the load force there are two paths where the velocity influences the load force dynamics: one through $A_{e\oslash}$ and another through the viscous friction B in the hydraulic cylinder.

Fig. 7.51 Force loops with natural velocity, with/without actuator viscous friction [17]

The force loop is designed with a velocity feedforward component, which effectively cancels the gain limiting zero in the force loop. Feedforward compensation is derived neglecting the acceleration term (Fig. 7.52).

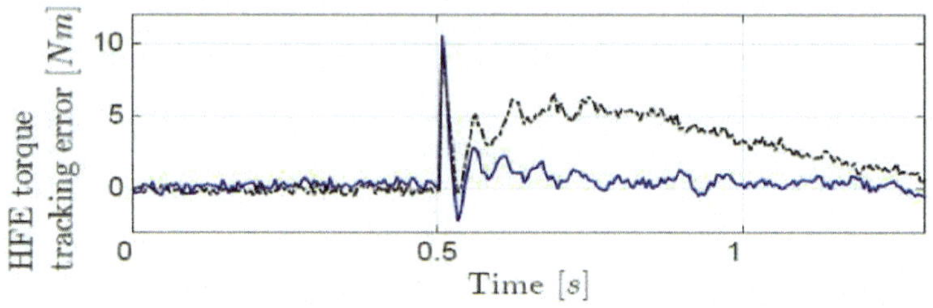

Experimental result for a 10 Nm torque step in the HFE joint: in dashed black the non-compensated torque tracking error ($RMSE$ = 2.34 Nm), which uses only a PID controller, and in solid blue the compensated one ($RMSE$ = 1.13 Nm), which uses the feed forward command u_{vc} to enhance the PID control response. The velocity compensation made the system about 3 times faster and reduced the RMSE by about 50%.

Fig. 7.52 Effect of velocity FF compensation [17]

Feedforward compensation reduces the force/torque error of the joint and provides a faster response.

The desired impedance is set in joint space using PD with an inverse dynamics controller, a model-based technology (Fig. 7.53).

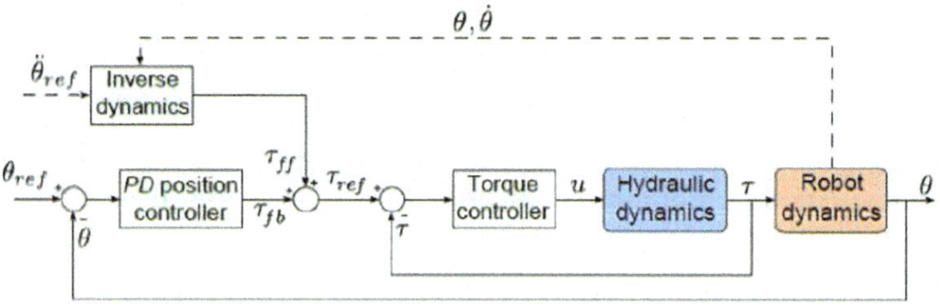

Block diagram for the HyQ cascade control with an outer feedback joint-space position loop. It creates, together with a feed forward term from the inverse dynamics, the torque reference τ_{ref} for the inner torque loop. The model-based inverse dynamics controller needs the feedback of the joint position θ and velocity $\dot{\theta}$, and also the reference acceleration $\ddot{\theta}_{ref}$ to calculate the feed forward command τ_{ff}.

Fig. 7.53 Desired force setting [17]

Experimental verification of control is performed by checking the force and position for 5 Hz sine input to the joint [18] also presents experimentally, control by comparison, matching passive compliance with virtual compliance (Figs. 7.54 and 7.55).

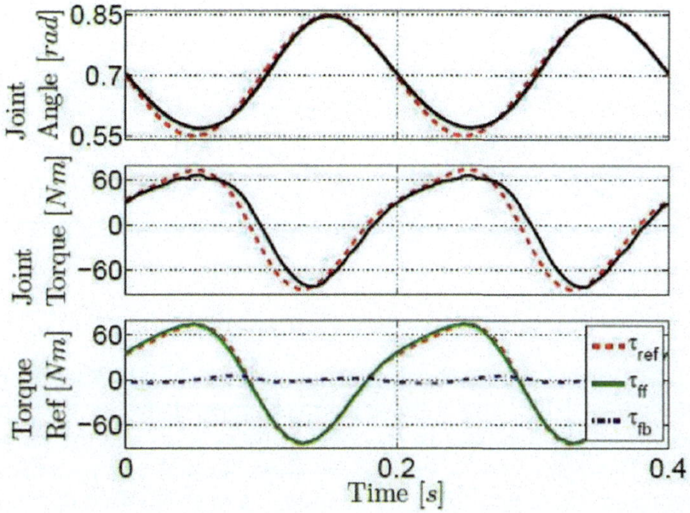

Experimental results showing a precise position and force tracking for a 5 Hz sine position reference for the HFE joint. In the first two plots, the red dashed lines indicate the reference command and the black solid ones the actual value. The last plot shows that the reference torque τ_{ref} ($\tau_{ref} = \tau_{ff} + \tau_{fb}$) consists essentially of the feed-forward term τ_{ff}.

Fig. 7.54 Control verification [17]

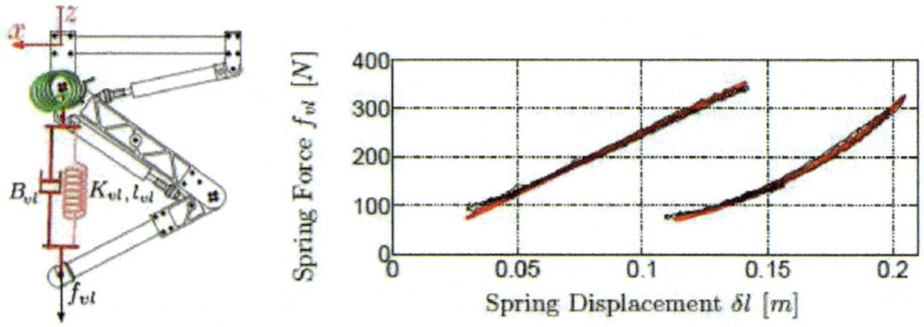

(a) (b)

Virtual components implemented on the HyQ leg: a rotational spring-damper at HFE, and a linear spring-damper between the hip and the foot. The virtual leg damping is B_{vl}, the stiffness K_{vl}, and the virtual spring length l_{vl}. The force f_{vl} is created by these virtual components. The red arrows represent the coordinates frame axes x and z. (b) Two different virtual springs implemented on the real hardware: a linear one ($f_{vl} = 2500\delta l_{vl}$, being $\delta l_{vl} = l_{vl_0} - l_{vl}$, where l_{vl_0} is the free length of the virtual spring) and an exponential one ($f_{vl} = 16\, e^{15\, \delta l_{vl}}$). The red line represents the desired stiffness profile, and the black line the one produced by the HyQ leg.

Fig. 7.55 Control efficacy to match passive compliance [17]

7.7.2.5 Control of 4 Axis, Half Axle Test Rig

Road load data simulation on a half axle, 4 actuator test rig is described in [19]. The system has 4 inputs: 2 force inputs in the x-and y-axes, position input in the z-axis and a moment about the y-axis and 6 outputs of 3 forces and 3 moments on the wheel. Measured road load data to be simulated are in terms of wheel forces/moments (Figs. 7.56, 7.57 and 7.58).

Fig. 7.56 $^1/_2$ axle, 4 actuator test rig [19]

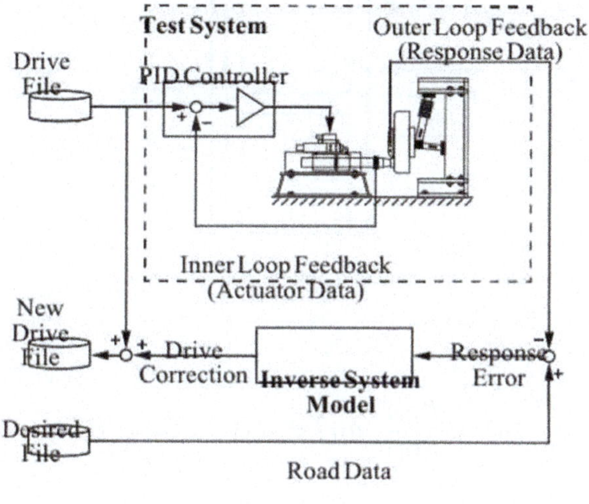

Fig. 7.57 Road load data simulation structure [19]

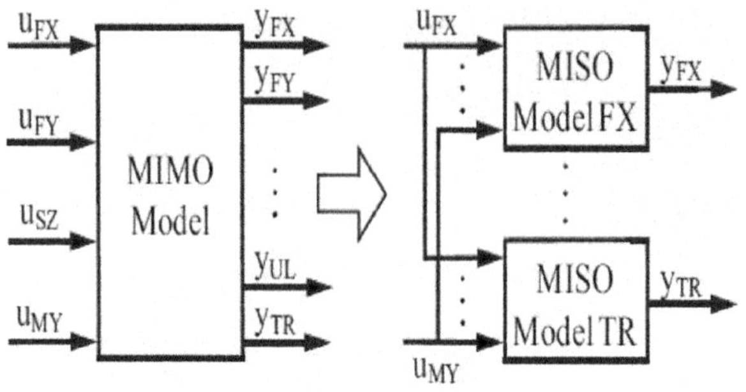

Fig. 7.58 MIMO model decomposition [19]

Suggested method:

- Decompose the MIMO model to 6 individual MISO models retaining all cross-couplings.
- Excite all channels individually by random white noise input recording all output responses.
- Identify for each channel the model of the system optimally by the model predictive method of system identification. The time domain model is transformed to Laplace form (Fig. 7.59).

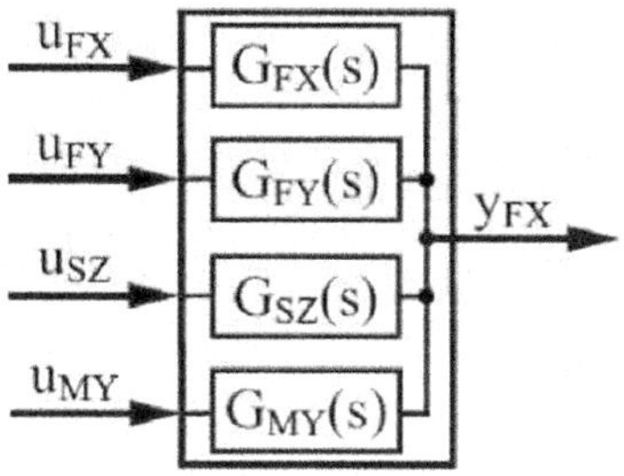

Fig. 7.59 MISO model. TIME domain/Laplace transformation [19]

$$y(t) = \frac{B(q)}{F(q)} \cdot u(t) \Leftrightarrow Y(s) = G(s) \cdot U(s) \tag{7.1}$$

- Compute the inverse of this model taking care of model order and structure.
- The force models have 4 zeros and 7 poles, while the inverse has 7 zeroes and 7 poles. Model structure is optimized by correlation study.
- All actuator drive signals are calculated by applying road load data to the computed MISO models. These drive signals are then applied to the test system.
- The procedure is iterated until satisfactory simulation is obtained (Fig. 7.60).

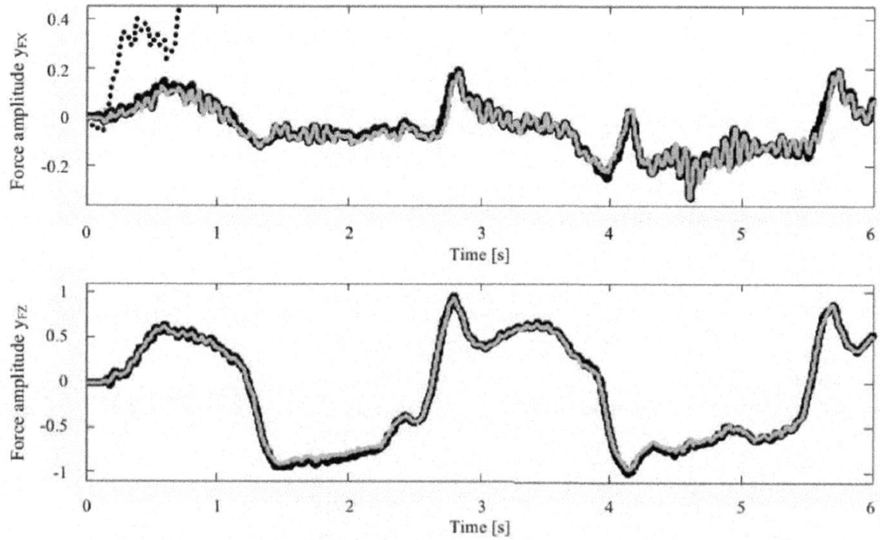

Comparison of road measured forces (black, solid) and laboratory simulated forces with structure optimization (gray solid) and without structure optimization (gray dotted) for system outputs y_{FX} and y_{FZ} (values scaled by the maximum of y_{FZ}).

Fig. 7.60 Control effectiveness [19]

7.8 Control of Parallel Manipulators

7.8.1 Cascade Control

Kinedyn, a sea-state simulator, is a 6DOF Stewart platform that is hydraulically powered.

The 30 ton motion simulator is powered by 6 hydraulic tandem actuators, each 6.2 m long, driving a 5 m × 5 m moveable platform. The use of a tandem actuator for counterbalancing and servo proportional valve flow rated at a 10 bar pressure drop sharply reduced the control power. The total power was limited to 1000 kw (Fig. 7.61).

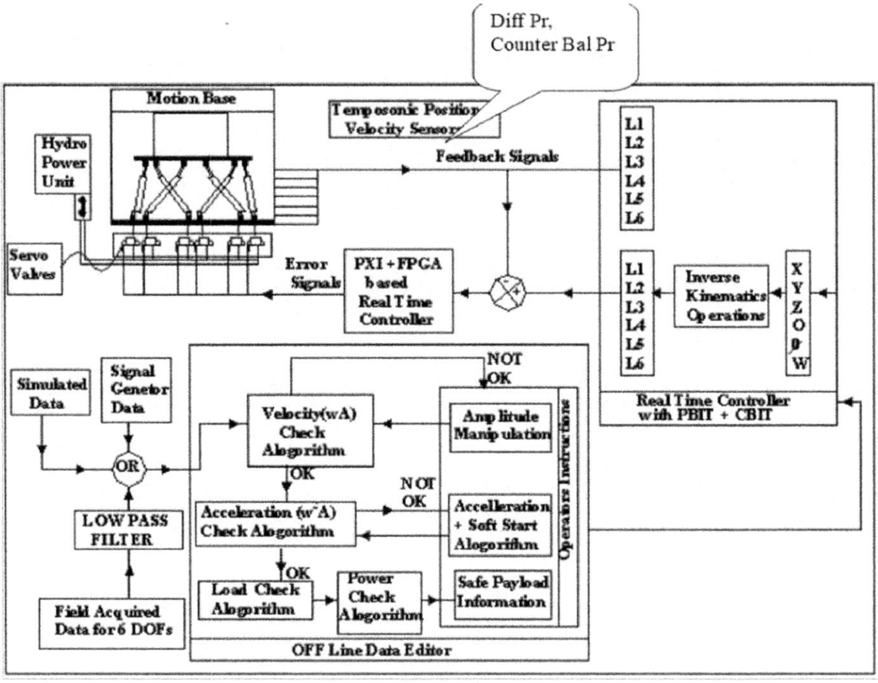

Fig. 7.61 Overall control scheme of Kinedyn [1]

The input trajectory is verified for exceedance of position, velocity and acceleration, offline, on kinematic and dynamic models of Kinedyn and wash out filters. Control has a built-in soft start feature that gradually raises the inputs from 0 to 1 in 5 s.

The Simulink implementation of position and force cascaded control is shown in Fig. 7.62. The cartesian commands are transformed to joint commands through inverse kinematics, which includes the feedforward components to the cascade controller.

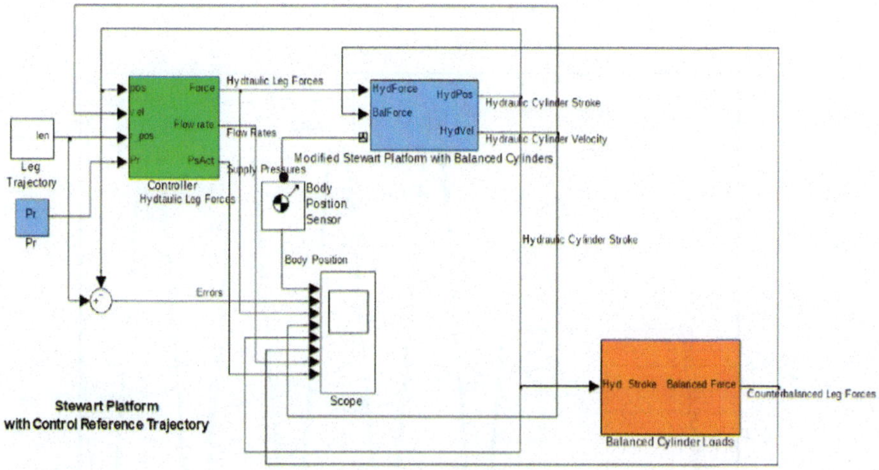

Fig. 7.62 Simulink implementation of cascade control [1]

A combined movement in cartesian space is shown in Fig. 7.63.

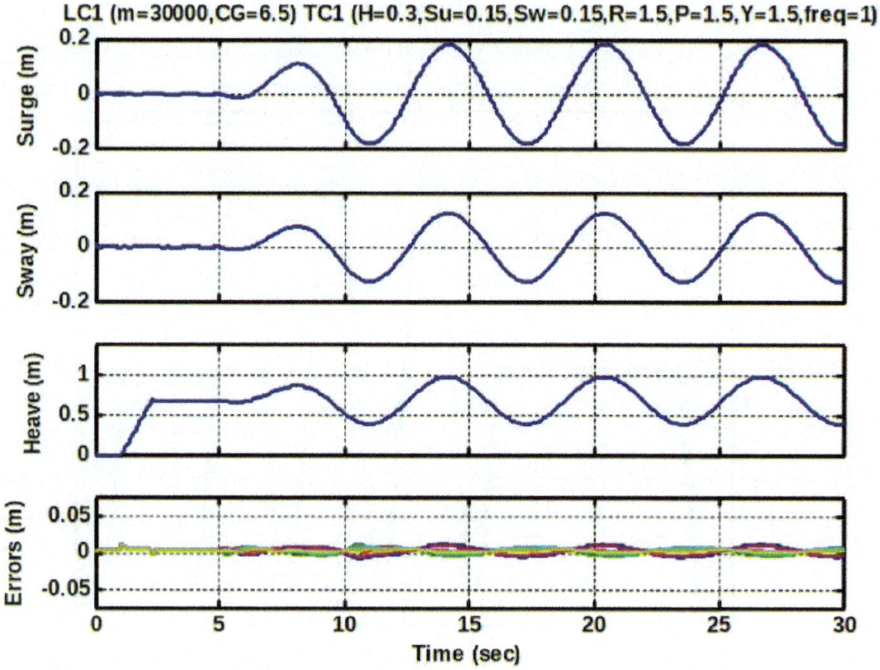

Fig. 7.63 Displacements in closed loop with soft start [1]

Robustness was studied for ±50% variation of assumed top mass, actuator damping, static load on actuator, actuator oil stiffness, and trapped oil volume, two of which are shown (Figs. 7.64 and 7.65).

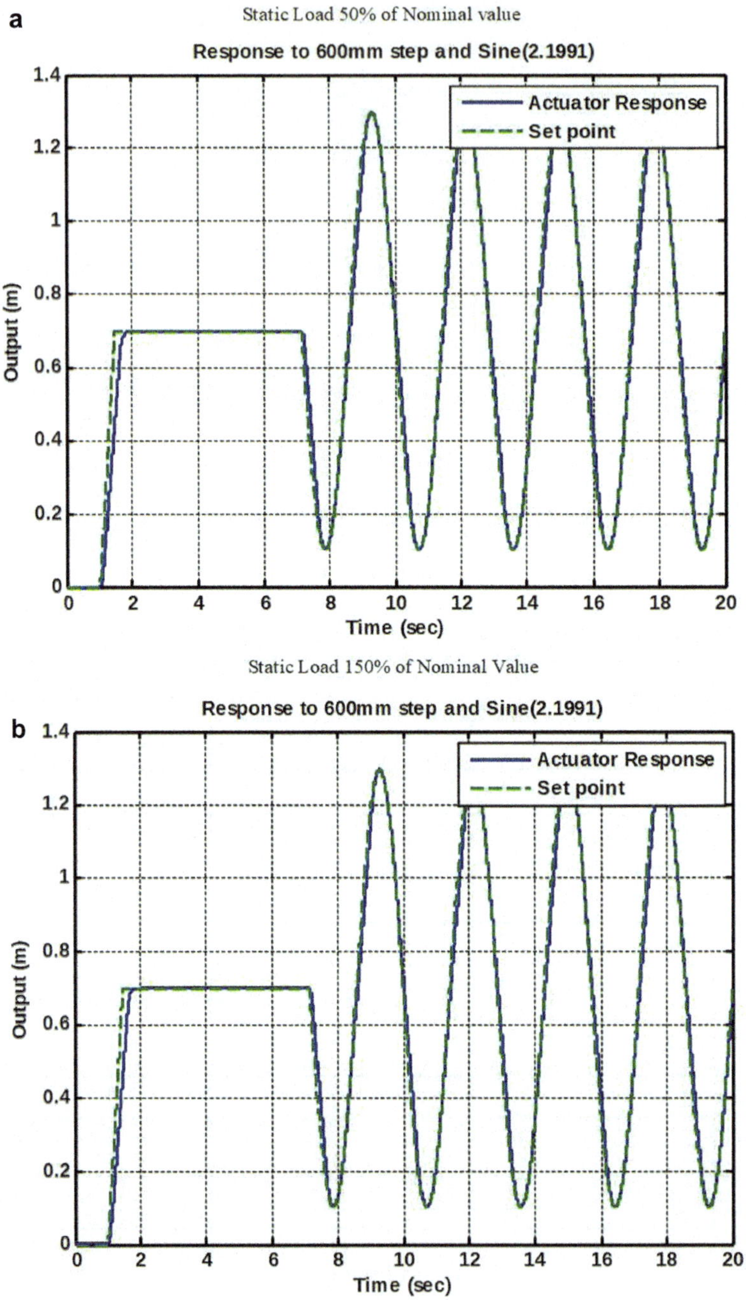

Fig. 7.64 **a** Robustness study for static load variation [1], **b** Robustness study for static load variation

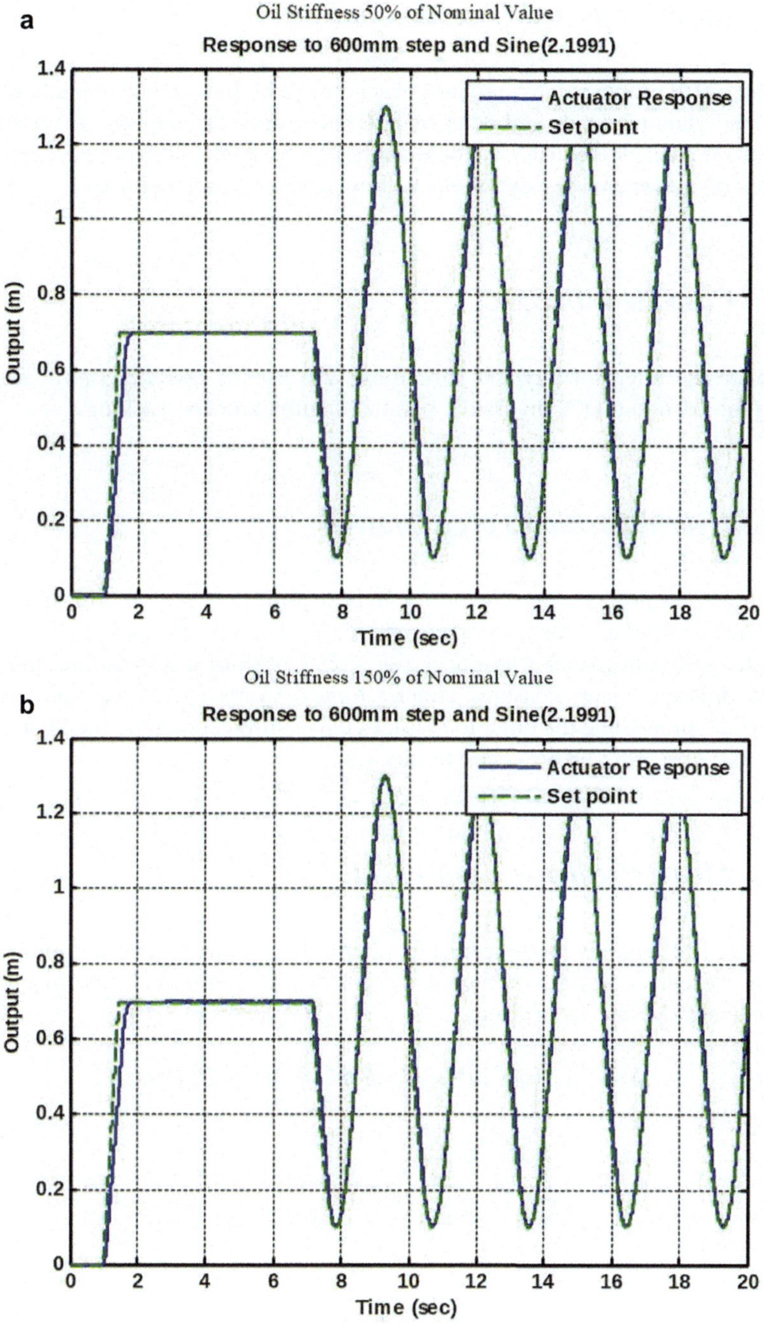

Fig. 7.65 **a** Robustness study. Oil compressibility 50% [1], **b**. Robustness studyoil compressibility 150% [1]

7.8.2 Observer-Based Cascade Control

Reference [19] proposes a disturbance observer to a hydraulic 6 DOF manipulator to assess uncertain payloads and other uncertainties, such as coupling, to compensate for them. A cascade controller is then applied. The approach avoids dynamics and complex direct kinematics and works without acceleration feedback.

7.8.3 Cascade with SMC

To reduce the effect of payload variations, [20] uses a cascade control based on sliding mode for a 6 DOF hydraulic parallel manipulator for tracking.

7.8.4 Hybrid Position/Force Control

Reference [21] proposed a scheme for hybrid position/force control of a 6 DOF hydraulic manipulator for a docking operation. The design uses a CCD camera and a range-finding laser transducer for global position information. An instrumented pressure plate measures contact forces. Control uses an inner loop for pose/orientation, while the outer loop tracks the position command based on contact forces. Experiments verify successful docking.

7.8.5 Model Predictive Control (MPC)

Reference [22] applies model predictive control for tracking and minimizing cost function based on an ARMA model. The inverse model includes an integrator to eliminate steady-state error (Figs. 7.66, 7.67, 7.68, 7.69 and 7.70).

$$S(q^{-1})\Delta(q^{-1})u(t) = -R(q^{-1})y(t) + T(q)w(t)$$

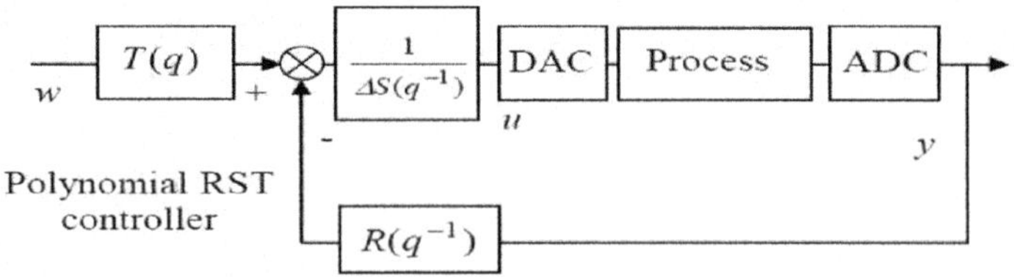

Fig. 7.66 Structure of MPC [22]

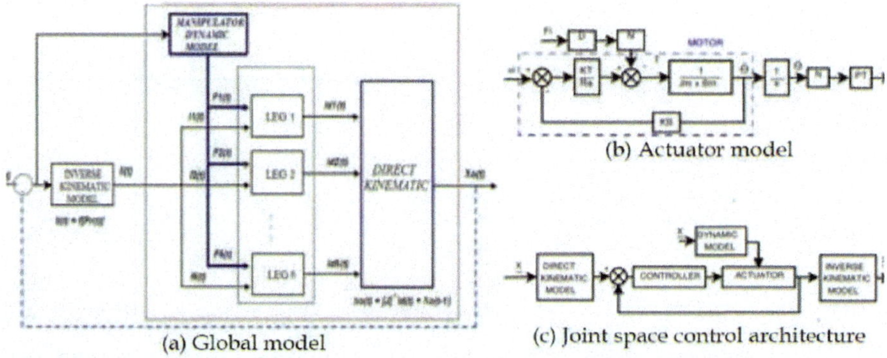

Fig. 7.67 Total system model [22]

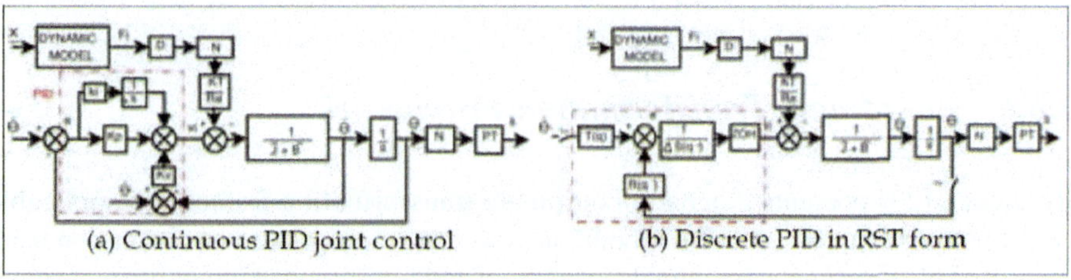

Fig. 7.68 Continuous/discrete models [22]

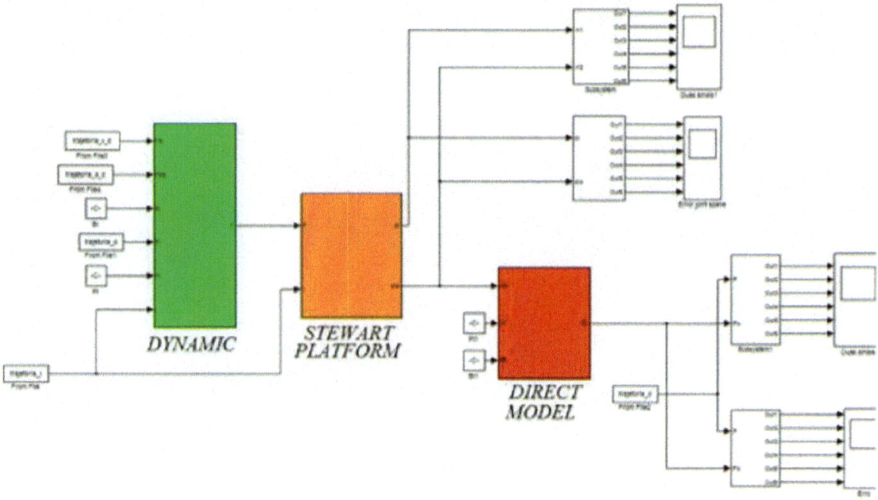

Fig. 7.69 Simulink dynamic and control model [22]

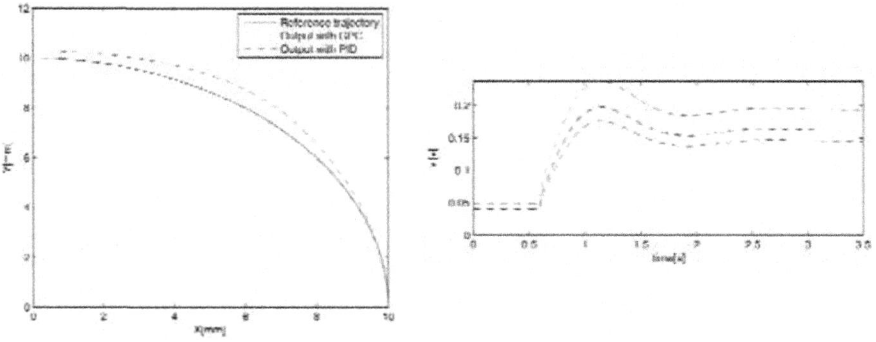

a) Tracking error b) Disturbance reaction, PID

Fig. 7.70 Simulation results [22]

7.8.6 Coordinate Transformation Framework

Reference [23] presents a general coordinate transformation framework applicable to multiaxis, redundant parallel kinematic systems. Appropriate transformation from measured position and force signals to position and force control inputs are presented. The application of this is illustrated for a 2-axis 1 DOF system, a full vehicle test, a cruciform test rig, and a multiaxis shaking table (Figs. 7.71, 7.72, 7.73 and 7.74).

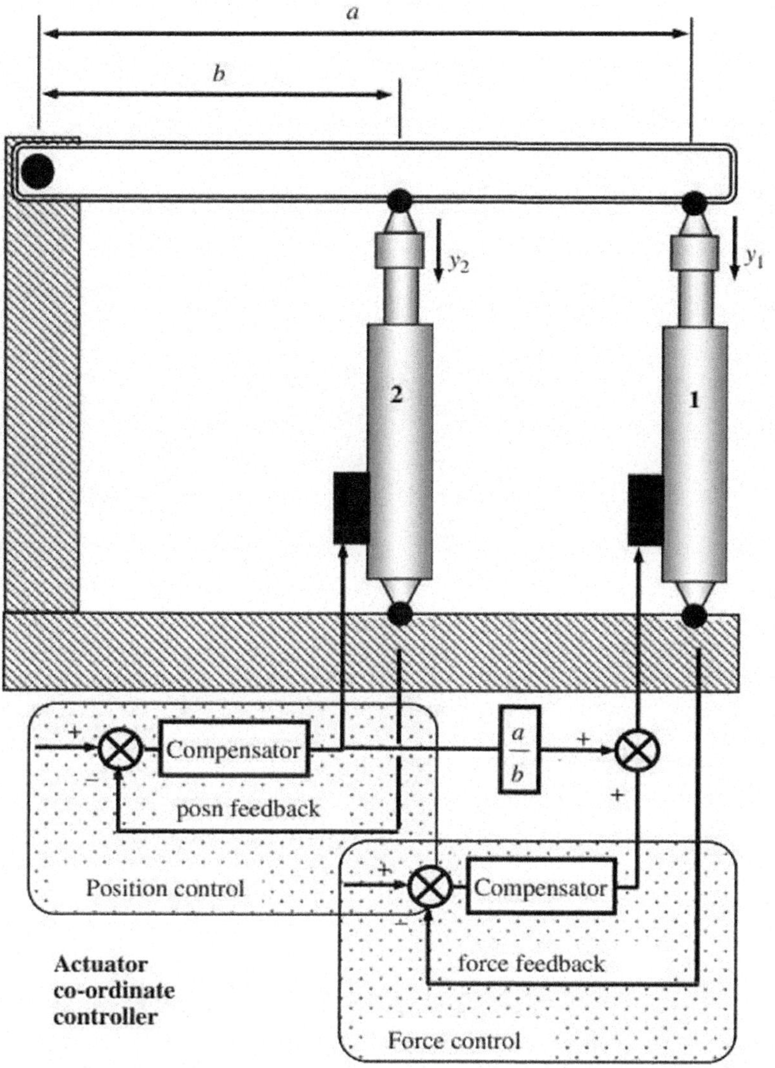

Fig. 7.71 2-axis 1 DOF system. The controller shown is in actuator coordinates [23]

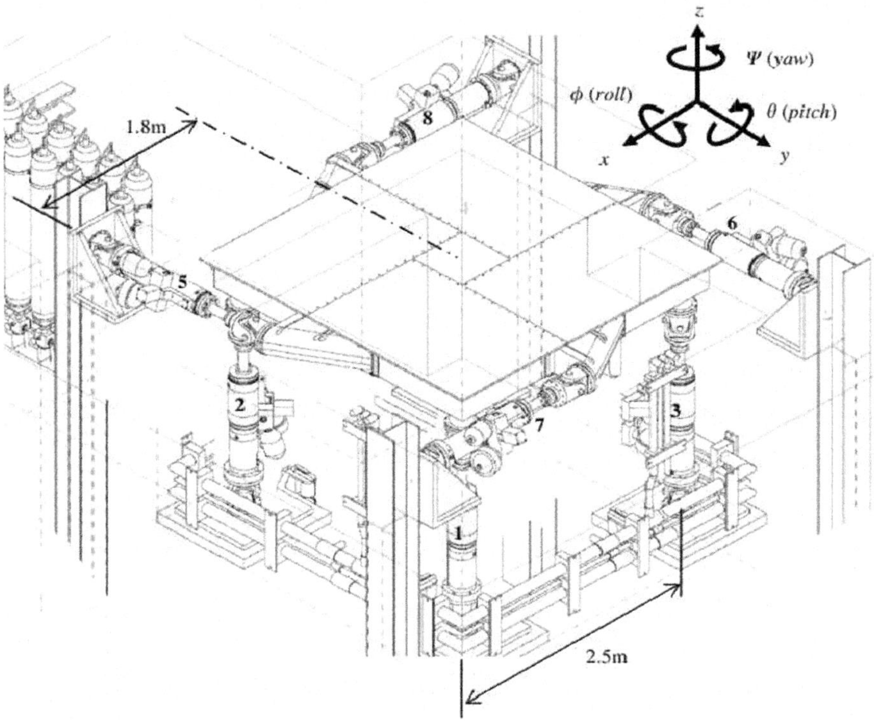

Fig. 7.72 6 DOF shaker table with 8 actuators [23]

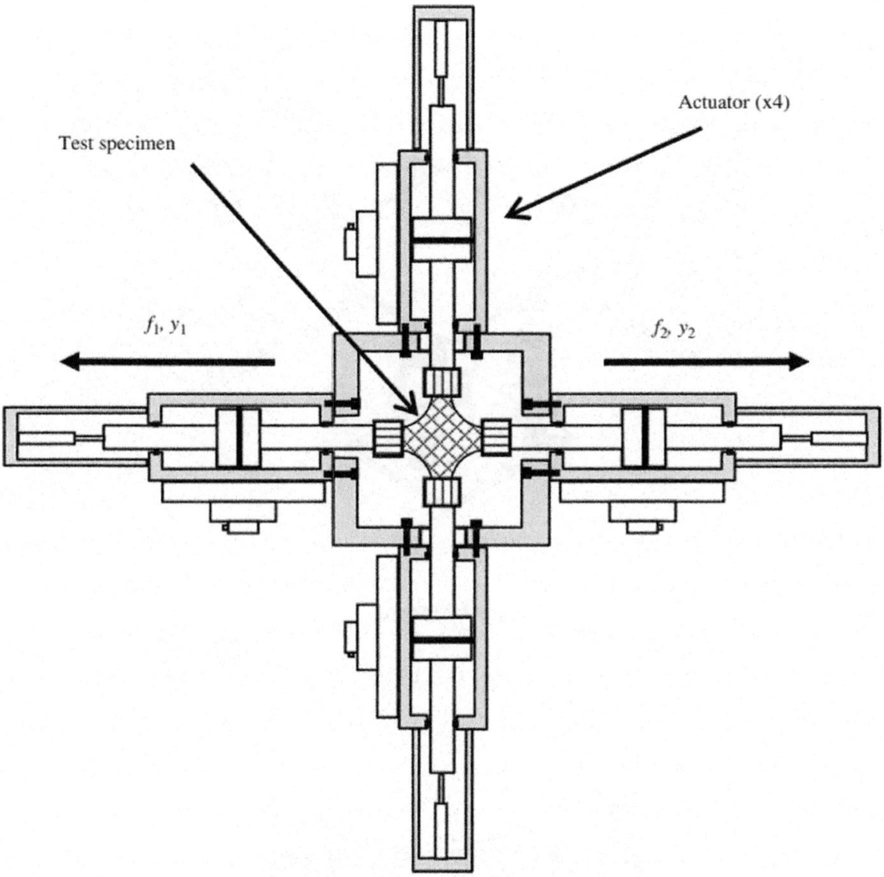

Fig. 7.73 Cruciform testing machine [23]

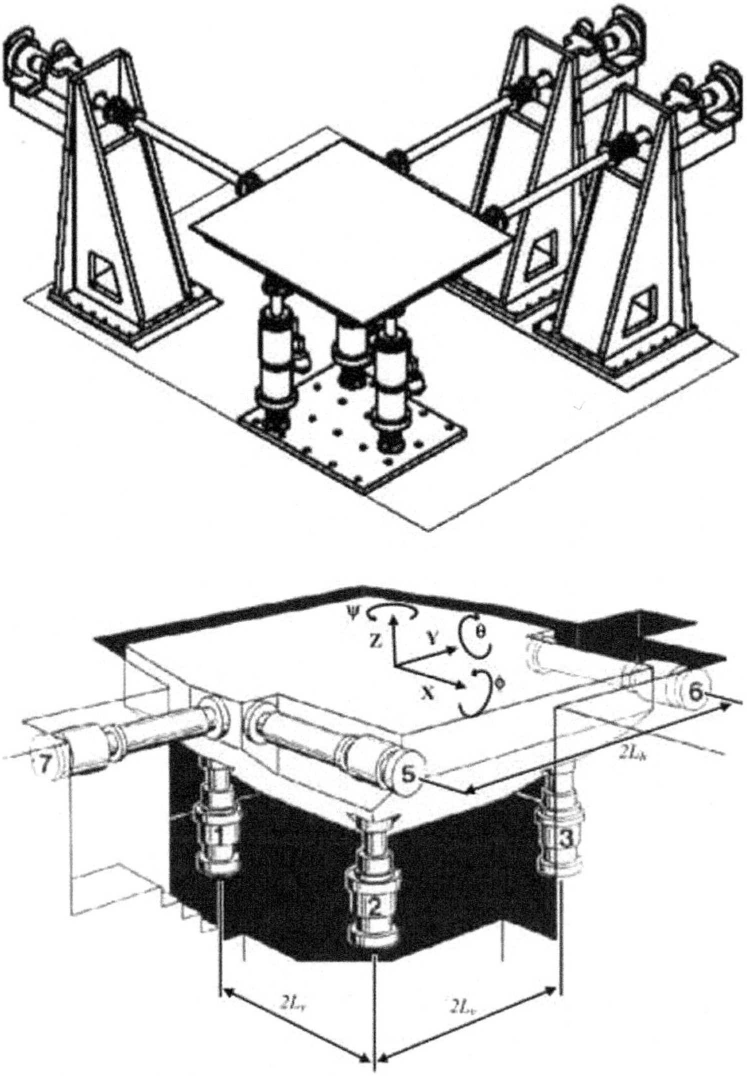

Fig. 7.74 A 6 DOF shaking table 0.5 ton [24]

7.9 Centralized Versus Decentralized MIMO Controllers

A basic distinction in MIMO controllers is between centralized and decentralized controllers. The methodology of controller design stems from FRF matrix measurement (Figs. 7.75 and 7.76).

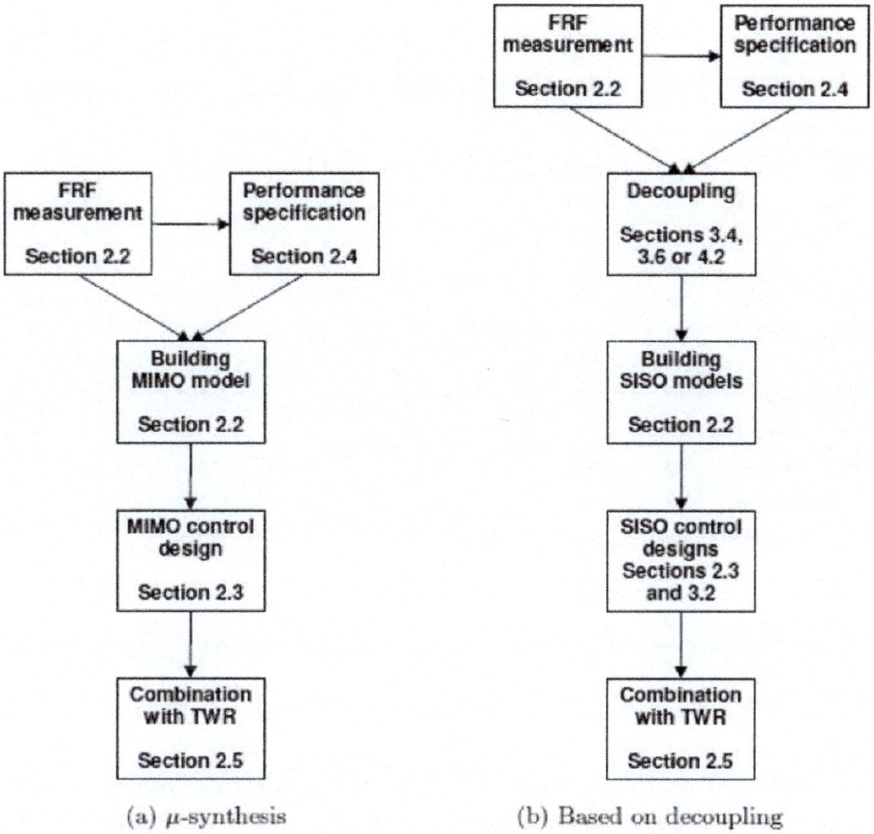

Fig. 7.75 MIMO controller design methodology [25]

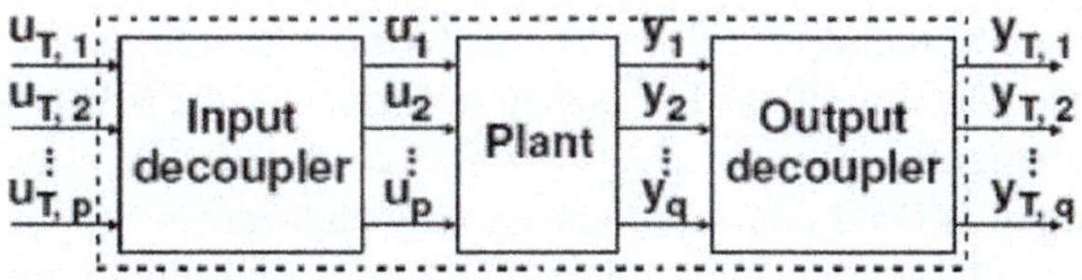

Fig. 7.76 Structure of decentralized controller [25]

The decentralized controller is step 2, based on the decoupler and individual SISO controllers [25] details and compares the two designs. FRF matrix measurement is common to both.

Sine or random or multisine excitations at different input levels are used for each of the inputs, and the FRF matrix is analyzed by measuring the output for each excitation.

Table. 7.4 Comparison of decoupling methods [25]

	MIMO model?	Integration with decentralized control design	Complexity (simple: 5, complex: 1)	Degrees of freedom (few: 1. many: 5)	Complete frequency range?	Accuracy (good: 5, bad: 1)
Align	No	With charact. loci	4	2	No	2
Multi-Iraq. align	No	With charact. loci	4	2	Yes	1
Optim..scaling	No	No	5	1	No	1
Approx. RFN	No	With charact. loci	4	2	No	2
DTM-based	No	No	4	2	No	3
Pseudo-diag	No	No	4	1	Yes	2
Simult. diag	No	No	4	4	Yes	2
Modal decoupl	No	No	3	2	Yes	3
Approx. frames	Yes	With charact. loci	1	5	Yes	1
Dynamic pseudo-diag	No	No	3	5	Yes	3
Smith-McMillan	Yes	No	1	5	Yes	1
Inverse-based	/	No	2	5	Yes	4

From the FRF matrix measurement, a parametric model of the system is estimated either in the frequency domain by the nonlinear least squares method or in the time domain by the subspace method.

From ARMA models, static decoupling is performed in several methods. A comparison of these methods is shown in Table 7.4.

The authors propose merging decoupling with any of the known SISO design methods. They recommend H∞ design considering robustness.

The centralized design referred to as synthesis derives the MIMO controller from the FRF measurements.

The demands of MIMO control of automobile endurance

Test rigs for road load data simulation are met by combining with TWR which is an iterative procedure.

The two methods are applied and compared for three test requirements:

- A half car suspension system
- A car resting on a CUBE on a wheel and
- A tractor on 4-posters (Fig. 7.77).

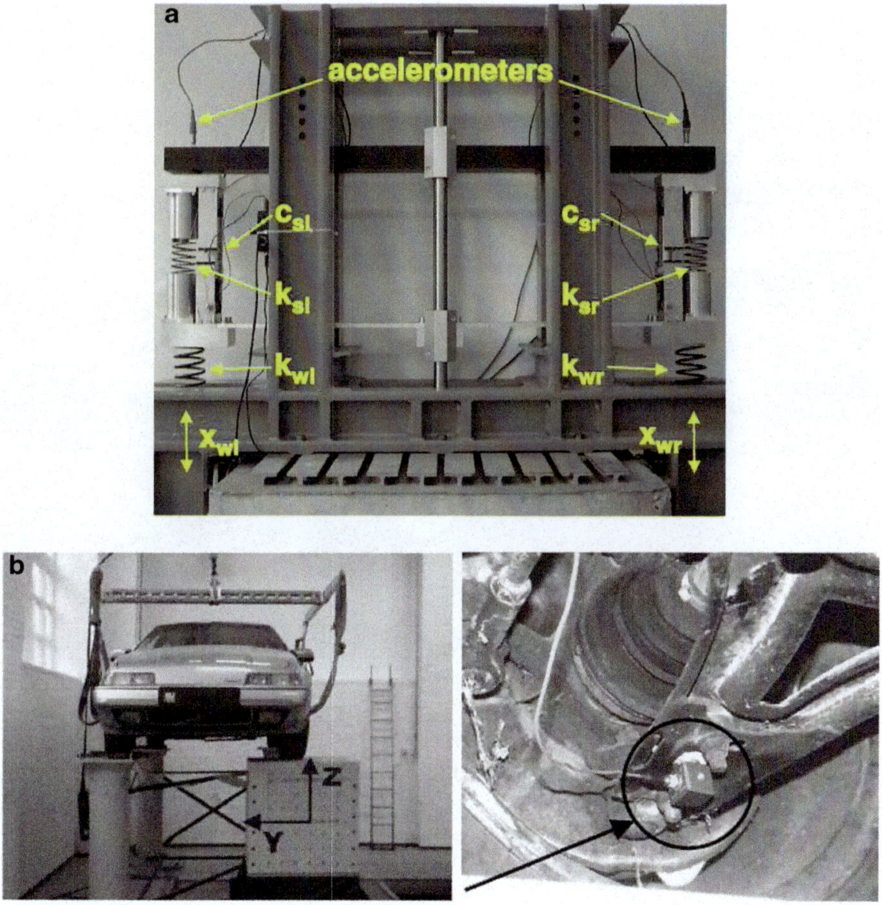

Fig. 7.77 **a** A half car suspension setup [25], **b** car with one wheel on CUBE [25], **c** tractor on a 4-posters [25]

c

Fig. 7.77 (continued)

A summary of the performance of the two controllers is presented in Table 2 (Table 7.5).

Table. 7.5 Summary of results [25]

	μ-synthesis		Decentralized		Dynamic decoupling		Static decoupling		Bandwidth limitations
	Order	rel. perf.	Order	rel. perf.	Order	rel. perf.	Order	rel. perf.	
Half Car	66 (47)	1	Unacceptable $\left(\text{predicted by } \mu_{\Delta_C}^{-1}(\boldsymbol{E})\right)$		$6 + 6(\mathrm{T_u}((s))$ $14 + 14(\mathrm{C}(s))$	1.02	$20 + 20$ $(\mathrm{C}(s))$	1.18	(a), (b), (c)
CUBE	114 (51)	1	$24 + 31$ (C(s))	1.52	$20 + 15(\mathrm{T_u}((s))$ $22 + 26(\mathrm{C}(s))$	0.94	Comparable to decentralized		(a), (b), (d)
Tractor	290	1	Unacceptable $\left(\text{predicted by } \mu_{\Delta_C}^{-1}(\boldsymbol{E})\right)$		98 $(\mathrm{T_u}(s))$ Unacceptable $\left(\text{predicted by } \mu_{\Delta_C}^{-1}(\boldsymbol{E})\right)$		4×23 $(\mathrm{C}(s))$	0.99	(a), (b), (c), (e)

Summary of experimental validation. Legend of used bandwidth limitations: (a) due to model uncert;.ue to interaction $\mu_{\Delta_C}^{-1}(\boldsymbol{E})$, , (c) due to NMP-zeros, (d) due to input saturation, and (e) due to delay.

References

1. Raghavendra DR (2010) Kinedyn—design report for control systems
2. Li L (2015) Simulation and control of servohydraulic actuators for test applications. Doctoral thesis. Graz University
3. Liu L et al (2019) A review of industrial MIMO decoupling controls. Int J Control, Autom Syst. TCX2019
4. Quan L et al (2013) Decoupling control research on the test system of hydraulic drive unit of a quadruped robot based on the diagonal matrix method. Intell Control Autom Sci Res
5. Rodermond S (2006) Modeling, identification and control of a multi variable hydraulic servo system. Masters thesis. University of Eindhoven
6. Gao et al Y (2018) Robust Decoupling of mixed sensitivity controller design of looper control system for hot strip mill process. Adv Mech Eng 10(11)
7. Vaes et al D (2004) Belgium. Optimal decoupling for MIMO controller design with robust performance. In: Proceedings American control conference, 2004
8. Zhu C, Zhang H, Wang W, Li K, Liu W (2020) Robot control of hydraulic tracked vehicle driven system on quantitative feedback theory. Int J Distrib Sens Netw 16(2)
9. Brahmi A, Zhu WH et al (2016) Adaptive control of mobile manipulator robot based on virtual decomposition approach. In: Proceedings of 13th international conference on information in control, automation and robotics ICINCO, 2016
10. Lahera P et al (2015) Model based development of control systems for forestry Cranes. University of Sweden. J Control Sci Eng 2015, article ID 256951. HINDAWI
11. Focchi M, Gugliemino E, Semini C, Boaventura T, Yang Y, Caldwell DG (2010) Control of hydraulically operated Quadraped Robot leg. In: International conference on robotics and automation Alaska US
12. Cunha TB et al (2010) University of Italy. Gain scheduling control for hydraulically operated HyQ Robot leg. In: ABCM symposium series in mechatronics, vol 4
13. Louci F, Kachida S (2020) Optimization of sliding mode control with PID surface for robot manipulator by evolutionary algorithms. Article. Open Computer Science 2020:396–407
14. Mattila J, Koivumaki J, Caldwell DG, Semini C (2017) A survey on control of hydraulic robotic manipulators with projection to future trends. IEEE/ASME Trans Mechatron
15. Clegg AC (2000) Self tuning position and force control of a hydraulic manipulator. Doctoral thesis. Heriot Watt University
16. Wang Y, Wang X, Li C, Ruan J, An S (2022) Research on position/force hybrid control of an FFHD robot with EHS actuators. J Braz Soc Mech Sci Eng 44:211
17. Boaventura T (2015) Model based hydraulic impedance control of dynamic robots. IEEE Trans Robot
18. Muller T et al (2016) Optimization of inverse model identification for multi axial test rig control. In: MATEC web of conferences, vol 42, 01002
19. Pi Y, Wang Y (2010) Observer based cascade control of a 6 DOF parallel hydraulic manipulator in joint space coordinates. Mechatronics 20(6)
20. Guo H et al (2007) Hybrid position/force control of 6 DOF Hydraulic parallel manipulator on a sliding mode. Control Eng Pract 16(9):1055
21. Guo H et al (2016) Hybrid position/force control of 6 DOF hydraulic manipulators using force/vision. Ind Robot. ISSN 0143-991x
22. Rosario JM et al (2010) Supervision and control strategies of a 6 DOF parallel manipulator using a mechatronic approach. Ontario University. Researchgate./net/publications/22908920
23. Plummer A (2010) A general co-ordinate transformation framework for multi axis motion control with applications in testing industry. Control Eng Pract 18:598–607
24. Plummer A (2016) Model based motion control for multi axis shaking tables. Control Eng Pract 53:109–122
25. Vaes D et al (2005) Optimal static decoupling for multi variable control design. Ph.D. thesis. Katholieke University of Leuven. Belgium

Chapter 8
Fault-Tolerant Hydraulic Systems

Abstract Some of the research towards Fault Tolerant Systems aimed for complex and expensive systems are cited here. Notes on Fault avoidance through contamination control, model based methods for Fault Detection and Diagnosis are included. The Signal based Fault Diagnosis methods in the research domain are mentioned.

8.1 General

As systems grow in complexity and cost, the need arises to improve the productivity and uptime of the system. Fault tolerance is best achieved with fault avoidance, fault identification, and fault monitoring. Considering that the majority of hydraulic system failures are attributed to solid particulate contaminated oil, this note includes a mention on contamination control.

8.2 Contamination Control

Solid contaminants in oil may cause catastrophic failures, but are certain to cause performance deterioration of hydraulic elements due to wear of working elements. Contaminant sensitivity of different hydraulic elements has been studied, and recommendations made to tolerance level of particulate contamination of the elements [1] (Tables 8.1 and 8.2).

D. R. Raghavendra, *Electrohydraulic Servo Systems*, Springer Tracts in Mechanical Engineering, https://doi.org/10.1007/978-981-19-8065-7_8

Table 8.1 Contaminant sensitivity of hydraulic elements [1]

ISO code numbers			Type of system	Typical components	Sensitivity
23 /	21 /	17	Low-pressure systems with large clearances	Ram pumps	Low
20 /	18 /	15	Typical cleanliness of new hydraulic oil straight from the manufacturer	Flow control valves Cylinders	Average
			Low-pressure heavy industrial systems or applications where long-life is not critical		
19 /	17 /	14	General machinery and mobile systems	Gear pumps/motors	Important
			Medium pressure, medium capacity		
18 /	16 /	13	World Wide Fuel Charter cleanliness standard for diesel fuel delivered from the filling station nozzle High-quality reliable systems	Injector valve and high-pressure pumps/motors	Critical
			General machine requirements	Directional and pressure control valves	
17 /	15 /	12	Highly sophisticated systems and hydrostatic transmissions	Proportional valves	Critical
16 /	14 /	11	Performance servo and high-pressure long-life systems e.g., Aircraft machine tools, etc	Industrial servo valves	Critical
15/	13 /	09	Silt-sensitive control system with very high reliability	High performance servo valves	Super critical
			Laboratory or aerospace		

Table 8.2 Particle concentration coded [1]

Allocation of particle counts to cleanliness classes		
No. of particles/ml		Cleanliness class
Over	Up to	
2,500,000		>28
1,300,000	2,500,000	28
640,000	1,300,000	27
320,000	640,000	26
160,000	320,000	25
80,000	160,000	24
40,000	80,000	23
20,000	40,000	22
10,000	20,000	21
5000	10,000	20
2500	5000	19
1300	**2500**	**18**
640	1300	17
320	640	16
160	**320**	**15**
80	160	14
40	80	13
20	40	12
10	**20**	**11**
5	10	10
2.5	5	9
1.3	2.5	8

ISO code numbers indicate the number of particles of size 4, 6, and 14 microns per ml, and the codes are defined in ISO 4059.

Since contaminant levels are specified in other standards, comparison/equivalence is shown in Table 8.3.

Table 8.3 Comparison of standards [1]

ISO/DIS 4406 BS 5540/4 codes	Defence Std. 05/42		NAS 1638	SAE 749
	Table A	Table B		
13/11/08			2	
14/12/09			3	0
15/13/10			4	1
16/14/09		400F		
16/14/11			5	2
17/15/09	400			
17/15/10		800F		
17/15/12			6	3
18/16/10	800			
18/16/11		1300F		
18/16/13			7	4
19/17/11	1300	2000		
19/17/14			8	5
20/18/12	2000			
20/18/13		4400F		
20/18/15			9	6
21/19/13	4400	6300F		
21/19/16			10	
22/20/13	6300			
22/20/17			11	
23/21/14	15,000			
23/21/18			12	
24/22/15	21,000			
25/23/17	100,000			

Systems have different contamination levels based on in-built and ingressed contamination through breathers and actuator rod ends. Based on these conditions, filters of adequate micron rating (Beta x > 100), flow and pressure ratings are placed in pressure and return lines of systems. In systems with open-circuit configuration, offline filtration units are the norm.

Currently, use of online monitoring sensors for contamination levels at pressure and return lines is common. These online sensors are located at bleed lines (Fig. 8.1).

	CS 1000	CS 2000

Brief description	- Contamination Sensor for measuring the solid particle contaminants in hydraulic and lubricating fluids	- Contamination Sensor for measuring the solid particle contaminants in hydraulic and lubricating fluids
Features	- Up to 300 bar - No data input - Switching output - Analog output: 4 – 20 mA or 0 – 10 V - RS485 port	- Max. 40 bar - 2 x 4 – 20 mA inputs for AquaSensor (AS) or other sensors - 3 dedicated and 3 optional outputs - RS232 for ISO code display - 2 alarm relays - Universal PLC output - Option: 4 – 20 mA, RS232, RS485 or Ethernet - Can be operated with an external flow sensor
Competitive advantages	- Recalibratable - IP 67 - Shock- and vibration-resistant - Compact design - 3-digit display - Data interfaces integrated in unit	- Recalibratable - Large number of data outputs - Wide range of hydraulic conditioning modules available - Electrical outputs
Applications	- Mobile hydraulics - Industrial facilities, e.g. machine tools, injection moulding equipment, presses, etc. - Filter carts - Scheduled maintenance	- Test rigs

Fig. 8.1 Online monitoring sensors [1]

8.3 Fault Diagnosis

In a closed loop system, faults occur at any of the elements forming the loop. While some faults as at the input or output elements are additive in nature, faults in the gain elements are multiplicative. Fault diagnosis is rendered complicated in MIMO systems. Fault detection, diagnosis is a research challenge [2]. Two approaches are practiced for this: model-based and signal-based methods.

8.4 Model-Based FDD

Model-based FDD relies on knowledge of system model and detects system faults and or sensors. FDD receives system input and the sensors output, generates fault information with its algorithm. Fault detection is by parameter estimation, observer based, or parity based [2].

8.4.1 Parameter Estimation-Based FDD

Faults in hydraulic systems attributed to fluid contamination leading to increased friction, leakage, and component wear. With the system defined by state variable model, fault detection by parameter estimation is founded on minimization of controlled output error.

Sepasi [3] presents a study on a setup to simulate friction by pneumatic actuator loading and leakage by needle valves. The model is identified and verified by experimental measurements.

Online measurement of system variables, output position and actuator chamber pressures, are made and an *unscented* Kalman filter (UKF) is used to extract system variables. An extended Kalman filter (EKF) is a nonlinear model while an unscented Kalman filter (UKF) reduces the order of the state variable model. Each of the faults were simulated and repeated measurements to ensure repeatability of results; measured parameters are compared with established model parameters. Mean average errors evaluated by statistical processes.

8.4.2 Observer-Based FDD

Observer-based FDD methods reconstruct the output of the system from measurements and use estimated error as residual. Observers can be linear or nonlinear. A class of observers referred to as fault detector filters that produce residual signals with directional properties aid in isolation of fault. Garimella [4] presents the application of adaptive robust observers (AROs) to detect common hydraulic system faults. The study was made on a setup on the swing arm of a robot inertially loaded. The study uses a bank of AROs for sensor fault detection and a bank of adaptive robust state reconstruction filters for state faults.

A similar study on multiple fault detection on an EHSA is reported in [5]. The interacting multiple model (IMM) makes use of a number of models for each defect and is associated with filters running in parallel. The output from each filter includes the state estimate, covariance, and likelihood of calculation [5].

Studies on multiple fault detection: friction. Leakages on an EHSA setup using IMM strategy were done. After identifying the system, measurements were done on KF-IMM and SVSF-IMM to determine the operating mode probabilities. Authors claim that SVSF has a higher probability for fault detect.

Sun [6] is a recent study on the subject of multiple fault detection on an aircraft hydraulic servo actuator. Authors propose an updated IMM (UIMM), where the current pattern is calculated by merging knowledge of previous moments to improve accuracy of the estimate. This is done at the beginning of each iteration, and the last estimated information of probability, state estimation, and covariance is used to estimate initial value of each filter. Compared to IMM, study claims UIMM drastically reduces the number of models avoiding model explosion to identify multiple faults (Fig. 8.2).

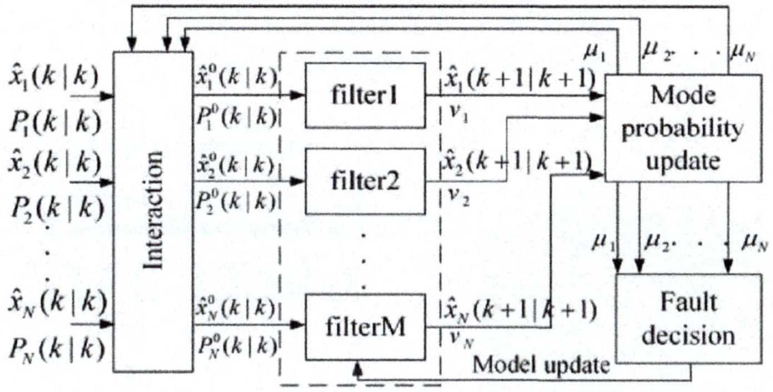

Fig. 8.2 Concept of UIMM [6]

8.4.3 Parity-Based FDD

Blisk manufacturing machine is a complex one using 11 actuators in 6 degrees. The critical linear friction welding process deploys an in-plane high-frequency oscillating drive with 2 no of 4-stage servo valves [7] (Fig. 8.3).

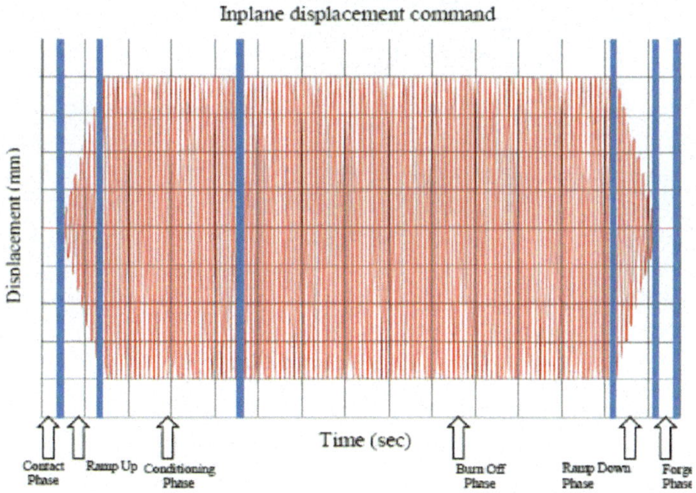

Fig. 8.3 LFW process in Blisk manufacture [7]

Faults in the drive can endanger uptime, and sometimes, even scrapping of components.

Fault detection is based on a model of the system which is compared with the actual system outputs to generate residual signal. Williams and Plummer [7] proposes the use of a Kalman filter for modeling which is fed with system inputs.

Residual generation is followed by evaluation and analyzed with respect to a domain expert system setting thresholds and logic for fault detection. Two types of fault identification are shown (Figs. 8.4 and 8.5).

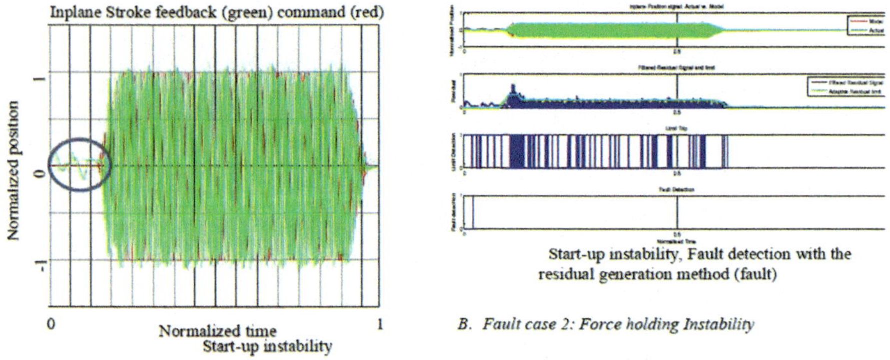

Fig. 8.4 Start up instability [6]

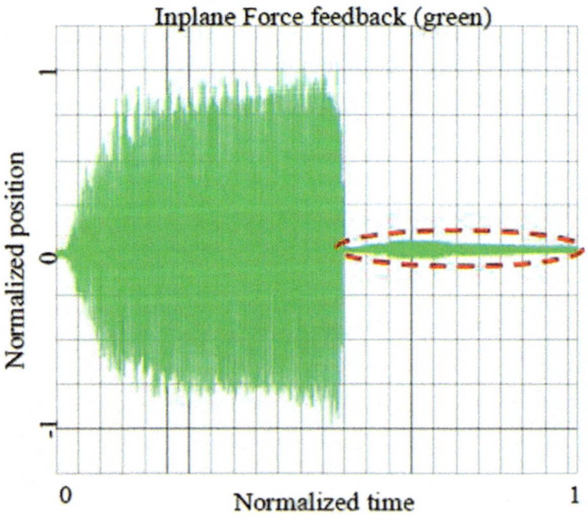

Fig. 8.5 Force hold instability [6]

8.5 Signal-Based Methods

Faults in hydraulic systems occur due to a combination of hydraulic and its structural interactions. Signal-based fault identification methods employ acoustic, vibration, temperature, and contaminant samples signatures [8] (Fig. 8.6).

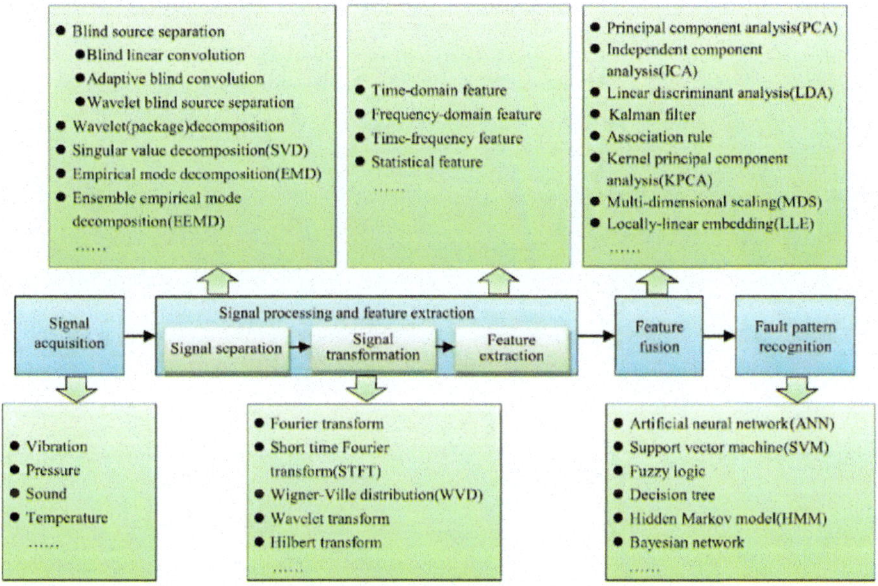

Fig. 8.6 Signal-based fault identification process [8]

Signal acquisition, signal separation, signal transformation, feature extraction, and subsequent feature fusion for fault pattern recognition, for which modern and intelligent approaches exist now.

In short, this is an extension of the familiar condition monitoring, acceleration signal processing done for machine bearing fault prediction.

References

1. Particle measurement Technology in practice. HYDAC.E7005–2–05–08-partik messtechnik
2. Rolf Isermann. Model based fault detection and diagnosis—status and applications. Darmstadt University Ph.D. Dissertation. 200Ad FAC.
3. Sepasi M (2005) Fault monitoring in hydraulic systems using Unscented Kalman Filter. University of Iran, Thesis for Master of Science
4. Garimella P (2005) An adaptive robust framework for model based fault detection. PhD Dissertation. Purdue University
5. Gadschen A et al (2011) Fault detection and diagnosis of an electro hydrostatic actuator using novel interacting multiple model approach. American Control Conference. San Francisco

6. Sun X et al (2020) Multi fault diagnosis approach based on updated interacting multiple model for aviation hydraulic actuator. Information published. Aug 2020. MDPI
7. Williams DT, Plummer A (2013) Fault detection of a linear friction welding production system using analytical models. In: 8th international conference on systems. ICONS
8. Dai J et al (2019) Signal based intelligent hydraulic fault diagnosis methods. Review and prospects. Chinese J Mech Eng 32–75. Open Access

The manufacturer's authorised representative in the EU is Springer Nature Customer Service Centre GmbH, Europaplatz 3, 69115 Heidelberg, Germany. If you have any concerns regarding our products, please contact ProductSafety@springernature.com

Printed and bound by CPI Group (UK) Ltd, Croydon, CR0 4YY
01/10/2026
02225870-0002